Physics for the *Utterly Confused*

Daniel Oman

Robert Oman

McGraw-Hill

New York San Francisco Washington, D.C. Auckland Bogotá
Caracas Lisbon London Madrid Mexico City Milan
Montreal New Delhi San Juan Singapore
Sydney Tokyo Toronto

Other books in the Utterly Confused Series include:

Financial Planning for the Utterly Confused, Fifth Edition
Job Hunting for the Utterly Confused
Calculus for the Utterly Confused

Library of Congress Cataloging-in-Publication Data

Oman, Daniel M.
 Physics for the utterly confused / Daniel M. Oman, Robert M. Oman.
  p.   cm.
 ISBN 0-07-048262-4
 1. Physics—Study and teaching.   I. Oman, Robert M.   II. Title.
QC30.066   1998
530—dc21   98-25808
                                                CIP

McGraw-Hill

A Division of The McGraw-Hill Companies

 3 4 5 6 7 8 9 0   FGR/FGR   9 0 3 2 1 0 9

ISBN 0-07-048262-4

*The sponsoring editor for this book was Barbara Gilson, the editing supervisor
was Stephen M. Smith, and the production supervisor was Pamela A. Pelton.*

Printed and bound by Quebecor/Fairfield.

McGraw-Hill books are available at special quantity discounts to use as premiums and sales promotions, or for use in corporate training programs. For more
information, please write to the Director of Special Sales, McGraw-Hill, 11
West 19th Street, New York, NY 10011. Or contact your local bookstore.

This book is printed on recycled, acid-free paper containing a
minimum of 50% recycled, de-inked fiber.

CONTENTS

How to Use This Book

Most Physics exams consist almost completely of problems. Unfortunately, many physics lectures and the majority of physics textbook space is devoted to *theory*. Theory is a four letter word to the non-physics major who is looking for the bottom line: What do I need to know for the exam?

This book explains thoroughly how to solve the most important types of problems in a non-calculus first year college physics course. This book would also be useful for an advanced high school course. Some physics help books toss thousands of problems at you with little explanation in between the lines. This book shows you how to solve the types of problems that most commonly appear on tests. **We spend a lot of time on each problem – even the simple ones – because we assume that you are having trouble even getting started with the simple problems.**

This book focuses on the methods that we have found to work for large groups of problems. If you develop the techniques we describe for solving problems then you will know how to successfully attack the problems you will encounter on the tests.

This book is designed for you to read through a chapter right before you do your homework problems. Then immediately start trying problems on your own. It is probably most efficient to go through one or two problems in this book and then find similar assigned homework problems and work them out.

You know you need this book if:

- you fall asleep during physics class
- when you try to read your textbook you catch yourself not concentrating and humming your favorite tune
- you are not particularly fond of math
- you want to get through physics in the most efficient way possible, without spending huge amounts of time on the course

Icons

The following icons are used throughout the book and can be used as bookmarks when reviewing the most important points in each chapter.

Remember

This icon highlights things that you should memorize. Right before a test, go over these items to keep them fresh in your mind.

Insight

This icon appears next to the "deeper" insights into a problem. If you have trouble understanding the details of why a problem makes physical sense, then this is the icon to follow.

Watch Out!

This icon highlights trouble spots and common traps that students encounter. If you are worried about making frustrating little mistakes or have an instructor that likes to try to trick you on tests, then this is the icon to follow.

Pattern

The intention of this icon is to you use the pattern of solving one problem to solve a general type of problem. In many cases the pattern is reviewed in a step by step summary and examples of similar problems and the general pattern of the solution are given.

Speed

Items next to this icon can be skipped if you are really struggling. On a second pass through the book, or for a more advanced student, this icon is intended to show you a few extra tricks, which will allow you to do problems faster. These items are included since speed is many times important to success on physics tests.

ADVICE TO THE *UTTERLY CONFUSED*

Many people believe the following: more work and more study results in higher grades. This is not necessarily so. You certainly must be willing to make a certain commitment of time and energy to this course but the key to academic success is concentrating your efforts on the right things at the right times. You may have noticed that those students who receive the highest grades are not necessarily the ones who work the longest number of hours. Some students may boast that they have studied all night for an exam, but don't be impressed by this habit. "Allnighters" and the like are almost always the result of procrastination and bad study habits. Getting no sleep before an exam is foolish and it usually takes several days to recover from this kind of nonsense. By taking advantage of the study techniques that follow you can achieve higher grades with less effort.

If you really want to get unconfused about physics, please take the following advice very seriously. You need to decide to develop some study habits that maximize your chances of surviving your physics course.

1. Do a quick reading on the topics to be covered in the lecture before attending class. Ten or fifteen minutes may be sufficient for a one hour lecture. The purpose here is to generally familiarize yourself with the topics to be discussed. Perhaps you can identify one or two questions or key points to listen for during the lecture.

2. Attend class and take notes. Attend all of the classes. That's right, we said **all** of the classes. Someone is paying for these classes so BE THERE! Be on the alert for any indication by the instructor of possible test questions. If the professor says something like "This is very important, you may be seeing this again," make a special note of this in your notebook.

3. This may be the most important step. **Do the homework problems regularly.** Regularly does not mean that you do the homework from chapter 1 two days before the test on chapters 1-5. In other courses it may be sufficient to read the text and review your notes, but in physics you must be able to work the problems. You don't learn problem solving skills by just reading examples of solved problems; you must do the problems yourself. By doing the homework problems on a regular basis you will be able to identify areas that you need more work on well in advance of the test. Physics problems can be difficult. Therefore, when you set out to work problems do not set yourself the task of working a certain number of problems but rather set out a certain amount of time to work on problems.

4. If you are having trouble understanding something – **ask**. Get to the bottom of your difficulties in understanding something. Ask the instructor, ask the teaching assistant, ask a classmate. Sometimes there is no substitute for being able to have a one on one conversation with someone about a question that you have. And don't be afraid to ask simple questions.

How to Prepare for a Physics Test

Examine the shelves of any bookstore catering to career oriented students and you will find books with titles such as: How to Pass the Real Estate Licensing Exam, or How to Succeed on the S.A.T. Examining these books will help you to understand how to take certain kinds of tests. One common thread in all books on how to pass particular exams is to know the questions in advance. Most writers of these types of books are in the business of training people in their particular areas, so they are close to those people that are making up the exams. This gives them the ready source of test questions, and knowing the questions (or at least the type of questions) is half way to knowing the answers. Therefore we make the following suggestions:

1. Almost all instructors in physics will place some problems on the test that are very similar to examples that they have done in class. Many times you may encounter the same problem with different numbers. This makes it very important to attend every class so as not to miss the opportunity to see possible test questions. If you do miss class, always get the notes from a friend.

2. Join a study group that does homework problems together. Sometimes slight modifications of homework problems appear on the test. A study group can be more efficient than grinding away on your own. Don't waste too much time with a study group unless it is productive. Your final preparations for a test should be done privately so that you can concentrate on developing a plan for taking the test.

3. Find sample physics tests given by your instructor for the past few years. It is a good bet that most of the questions for the exams in the near future will be very much like those of the immediate past.

4. Some physics problems involve mathematics that can be deceptively easy. For example, if you expect problems involving the manipulation of logarithms or exponents be sure you practice the mathematical operations and entering the numbers into your calculator so you don't have to stop and figure out how to take exponents during the test. Practice any unfamiliar mathematical operations before the test.

5. One of the common fallacies in preparing for exams is to prepare for the wrong thing. Many students will prepare for a physics exam by reading the text or by reading solutions to problems. But a physics exam is not a reading exam but a writing and problem-solving exam. If you have not **practiced writing solutions** to typical problems, you have not prepared as well as you might for the exam. Another advantage to this type of practice is that it increases your speed in writing solutions to types of problems that are likely to be on the test. This will allow you more time during the test to spend on unexpected or more troublesome problems.

Timing and the Use of the Subconscious

Have you ever experienced the frustration of having a conversation with someone and forgetting momentarily a name or fact that is very familiar to you? Usually, shortly after such an experience, the name or fact will come to you when you are not consciously trying to recall it. Another variation of this same phenomenon is when a person doesn't feel right about making a decision immediately upon receiving or defining a problem. They like to "sleep on it." Both of these situations have a common characteristic - the use of the subconscious. The fact that solutions are often presented to us in the absence of active work on the problem at the moment we receive the solution indicates that another part of the brain was analyzing the pertinent information and providing a solution. We call this part of the brain the subconscious, and this part of the brain is very effective at solving problems.

Here are some tips for effectively using the subconscious:

1. Your subconscious will not work without information. You must consciously sort out all of the facts or information for a particular problem. If you are having difficulty with a problem, try to get straight in your mind what you *do know* about the problem. Then also define in your mind what specifically you don't know or don't understand about the problem.

2. Put conscious effort into the problem up to the point of confusion. For some readers this may take only about 5 seconds, but the point here is that many people grind and grind on a problem for hours yet accomplish very little. It is more efficient for you to plan your study time so that you do not put yourself in a situation where your only choice is to grind on a problem.

3. After you have done all you can consciously on the problem, "Put it in the back of your mind." Don't keep worrying about it. It is important that you clear your mind so that you can accept the solution when it comes.

4. When a solution comes, be sure to act on it quickly, so you can go on to something else. Sometimes instead of a solution to the problem you will receive a request for more information. The problem may still be unanswered, but will be clearer to you. What could be happening here is that your subconscious has analyzed the problem and found an essential piece of information missing and is asking you for it.

Strategies to Use During a Physics Test

You are now entering the test room. You are well prepared to take the test. You have practiced writing out solutions to some of the problems and know what to expect on the exam. You have gotten a good night's sleep the night before and eaten a healthy breakfast that will provide you with the energy needed for good concentration. You have a positive attitude. At this point worrying about how you will do on the exam is useless. Study time is over. You now need to concentrate on the strategies that will get you the highest possible score on the test. Here are some suggestions:

1. It is usually a good idea to take a minute or two at the beginning of the exam to look over all the questions. Look for the type of questions that you expected and have practiced and do these first. Save the hardest questions for last. It can be very frustrating to run out of time working on question # 4 only to realize that you didn't even get a chance to start question #5, which was much easier.

2. Have a rough idea of how much time you should be spending on each question. Sometimes certain questions will count for more points than others and the instructor should provide that information on the test.

3. If you are required to memorize a lot of formulas you may want to take the time at the beginning of the test to write down a few of the more complicated ones next to problems that involve those formulas as you are glancing over the test. Later during the test, your mind may be cluttered with formulas and it may be harder to correctly recall one of the more complicated ones.

4. Always include the units of your answer (miles per hour if the answer is a velocity, for example). Don't make the mistake of not including units. This is very important to almost all physics teachers.

5. Write your work clearly when you are solving a problem. This makes it easier for the professor to give you partial credit if they can clearly see that you did the problem correctly and just missed a minus sign.

6. Think about your answer to a problem. Does the answer make sense? For example, if you are solving for the length of one side of a right triangle and you are given the hypotenuse, your answer better not be a length greater than the hypotenuse. It is very important to be able to think like this on a test. This will help you to catch a lot of mistakes like missing a minus sign.

7. Unfortunately some instructors give tests that are much too long for a given period of time. It seems as if they are more interested in measuring how **fast** you can do physics than how **well** you can do physics. Try to find out in advance of the test if your professor's tests are like this. If the cutoff for an A is usually 75% instead of 90% then you need to be aware of this. This will save you from panicking as you run out of time on the test. Remember that you may be able to work for partial credit on that last answer. On these kinds of tests it is very important to keep your cool and try to get as many points as you possibly can. Stay positive all the way through and give it your best shot!

8. Make sure you know the difference between radian mode and degree mode on your calculator when taking a test that includes trigonometry (see Chapter 1, Mathematical Background).

9. Avoid prolonged contact with other students immediately before the exam. Many times the nervous tension, frustration, defeatism, and perhaps wrong information expressed by colleagues can be harmful to your performance.

10. Multiple choice tests: Find out if there is any penalty for a wrong answer. If not, don't leave any question unanswered. Find out if there is any partial credit for showing your work on a separate sheet of paper. One thing to think about for multiple

choice tests is how the professor is generating the choices other than the correct answer. Here are some typical wrong choices on a multiple choice physics test:

a) A formula requires the input of length in meters. In the problem the length is specified in centimeters. The wrong answer is off by a factor of 100.

b) A formula requires the input of a radius. Diameter is given in the problem. The wrong answer is off by a factor of two.

c) A question asks for a velocity. Choice A is 10 lbs. This is the correct number, but the wrong units. Choice D is 10 miles per hour, the correct answer. The lesson here is to look carefully at all the choices.

Your Self-Image as a Student

To a large extent, many people perform at the level of their own self image. One thing to get straight in your mind at the beginning of the course is that you are capable of mastering the material in your physics course. Some students get stuck in the mode of saying something like, "I have always been a C student." Perhaps we can convince you that that doesn't necessarily have to be true. In a series of three to five sequential mathematics courses, for example, it is virtually impossible to go from one course to the next, let alone a sequence of several, without eventually mastering the material in each previous course. Think back to your first math course where you were taught how to add, subtract, multiply, and divide. At some point in that course you may have thought that you *couldn't* understand certain concepts. By now you have mastered those skills. College physics is the same way. You are mentally capable of understanding and even mastering basic physics. Now it is true that different people learn at different speeds. If this is the case you may need to spend a little extra time or, more likely, make more effective use of your time.

Please remember that all of the study techniques outlined in this chapter are designed to make achieving higher grades *easier* for you. The sooner you become more organized and focused on your goals, the sooner you will begin to realize that you are capable of impressive accomplishments with a reasonable amount of effort.

Perhaps physics is a favorite area of study (but just a little confusing right now) and you may wish to pursue it in the future. Or, more likely, perhaps you are primarily interested in the most efficient way to make it through this course. Whatever you choose for your major area of study, find something you enjoy and pursue excellence. Give it your best today, and better tomorrow. We wish you success.

1

MATHEMATICAL BACKGROUND

Units

- Learn the metric terminology denoting powers of ten. Examples: micron or micrometer both refer to 10^{-6} meters, a nanometer is 10^{-9} meters. Familiarize yourself with the prefix table in the textbook. Don't get caught on a test not understanding a unit or its prefix.
- Get into the habit of **always labeling the units of your answers!** Physics teachers are notorious for deducting points even if you have the right number for the problem, but have not labeled the units.

Remember

One important concept to understand is that you can convert the units of a number by multiplying by fractions that are equal to one. Since 5280 feet = 1 mile then multiplying something by $\dfrac{5280\,\text{feet}}{1\,\text{mile}}$ is just multiplying by one.

1-1 A jet is travelling at 600 miles per hour. What is its speed in meters per second?

Solution:

$$\frac{600\,\text{miles}}{\text{hour}} \times \frac{5280\,\text{feet}}{1\,\text{mile}} \times \frac{1\,\text{meter}}{3.28\,\text{feet}} \times \frac{1\,\text{hour}}{60\,\text{min}} \times \frac{1\,\text{min}}{60\,\text{seconds}} = 268.3\ \text{meters/sec}$$

First we made the top unit read meters and then concentrated on the bottom unit so that it would read seconds.

1-2 For the formula $x = (1/2)\,at^2$ suppose that x has dimensions of length and t has dimensions of time. What are the dimensions of a?

Solution: The factor of $1/2$ is not relevant to the question of what the dimensions of the answer are. What we have is $\text{length} = a \times \text{time}^2$, therefore $a = \text{length/time}^2$.

1-3 Suppose you have spent 15 minutes on a test trying to solve a problem and have come up with the following formula. Time remaining in the test is short and you are about to start plugging in numbers for variables to solve for the answer. But hold on a second, is the following formula really correct?

Watch Out!

$$\text{Formula: } x = (1/2)\, at^2 + (2/3)at$$

Solution: The answer is no! Just by looking at this equation you should be able to see that the dimensions of the first term in the equation are not the same as the dimensions of the second term. For the first term, the dimension is length, while the dimensions in the second term (everything after the plus sign) are length/time. We cannot add quantities that have different dimensions and get a single answer. If you are not careful and simply plug numbers in for the variables in the above equation without keeping track of the units, you will end up with the wrong answer. Noticing the mistake in the dimensions may allow you to look back in the problem and catch the mistake. Maybe you forgot to square the variable t at an earlier point in the problem.

Suppose you were asked, "What is 10 meters plus 20 pounds?" The question doesn't make sense, does it? Adding the two terms together in the mathematical equation above is just like trying to answer this question by saying "30". The number 30 is totally meaningless in this context, 30 what?

Here's another simple example: What is 33 feet plus 10 yards? The answer is 21, right? No! It is in your best interest right now to get into the habit of always writing the units of your answer. 21 is the correct number for the answer, but many physics teachers will deduct points for not giving the correct units. And if enough time passes, will you really remember when you glance back at your test paper where the number 21 is written whether is was 21 feet or 21 yards that you were talking about back up there? Just for the record it was 21 yards.

Trigonometry

Memorize this:

$$\sin \theta = o/h$$
$$\cos \theta = a/h$$
$$\tan \theta = o/a$$

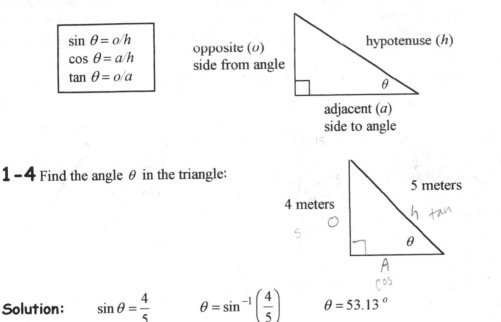

opposite (*o*)
side from angle

hypotenuse (*h*)

adjacent (*a*)
side to angle

Remember

1-4 Find the angle θ in the triangle:

5 meters

4 meters

Solution: $\sin \theta = \dfrac{4}{5}$ $\theta = \sin^{-1}\left(\dfrac{4}{5}\right)$ $\theta = 53.13^{\,o}$

The notation $\sin^{-1}(4/5)$ can be confusing. It **does not** mean $1/\sin(4/5)$ as you might initially think. The term $\sin^{-1}$ is read "inverse sine" and is a separate function on your calculator.

Make sure your calculator is in degree mode and not radian mode! Take the time now to understand how your calculator switches modes. If you haven't figured this out yet, go get your calculator now. Find the "mode" button and switch from deg to rad. Now take the inverse sine of 4/5. Did you get 0.927? This means 0.927 radians. To convert from radians to degrees recall that there are 360° in a circle and that the circumference is $2\pi r$, where r is the radius.

Watch Out!

Remember this conversion: $2\pi \text{ radians} = 360^o$ or $\dfrac{2\pi \text{ radians}}{360^o} = 1$

Remember

Now convert back to degrees: $0.927 \text{ radians} \times \dfrac{360^o}{2\pi \text{ radians}} = 53.13^o$

This is the answer in degrees that we got originally when the calculator was in degree mode. It would be a good idea now to reset your calculator to degree mode. That's usually where you'll be using it on a test.

Speed

For those of you who would like to see some fast ways to calculate sides and angles for certain triangles, here are a couple things to remember. The Pythagorean theorem is used to calculate the sides of a right triangle (a triangle with a right angle or 90° angle). The equation is $a^2 + b^2 = c^2$ (a and b are sides and c is the hypotenuse). For the triangle in problem 1-4, what is the length of the third side? The answer is 3 meters. Notice that $3^2 + 4^2 = 5^2$ or $9 + 16 = 25$. It is also true that $6^2 + 8^2 = 10^2$.

Another triangle to remember is the right isosceles triangle.

From this traingle notice that

$$\sin 45^o = \cos 45^o = 1/\sqrt{2} = .707$$

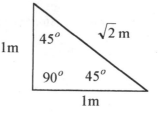

Vectors

A **scalar** quantity can be described by a single number. Examples are a boy with a mass of $100\,\text{kg}$ or a glass of water with a temperature of $18^o\,\text{C}$. Other quantities such as displacement, velocity, or force have a direction associated with them and are called **vectors**. In many texts vectors are written in *bold italics* and scalars in plain type or *italics*.

A vector is usually shown as an arrow oriented in space with the length (of the arrow) representing the number, and the orientation the direction. A vector can be described with a number and an angle as $A = 23\angle 37^o$. A vector is also commonly described using compass directions. This can be confusing to the first-timer.

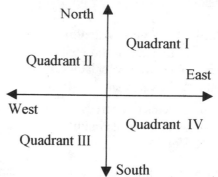

One type of problem that you will encounter many times is the **addition of vectors**.

1-5 Find the magnitude (number of Newtons of force) and direction (angle from 0 to 360 degrees) for the Resultant of the two forces: $F_1 = 100\,N$ at an angle of 40° north of west and $F_2 = 200\,N$ at an angle of 30° west of south.

Solution: Step 1. Draw a diagram of the problem.   Label the forces and their components.

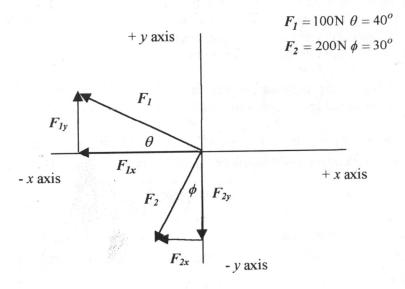

$F_1 = 100N \ \theta = 40^o$

$F_2 = 200N \ \phi = 30^o$

Before going any further, can you visualize the direction of the Resultant? Think of two forces pulling on an object. One force pulls up and to the left. A force twice as strong pulls down and to the left. The Resultant will be in the third quadrant, or an object pulled by these two forces will move down and to the left. **Insight**

Step 2. Find the magnitude of the components using trigonometry.

Refer to the diagram above and follow along with your calculator on this example to make sure that you know how to handle all of the trigonometric functions, including inverse trigonometric functions.

$$\sin\theta = \frac{o}{h} = \frac{F_{1y}}{F_1}; \quad F_{1y} = F_1 \sin 40^o = 100\,N \times 0.643 = 64.3\,N$$

$$\cos\theta = \frac{a}{h} = \frac{F_{1x}}{F_1}; \quad F_{1x} = F_1 \cos 40^o = 100\,N \times 0.766 = 76.6\,N$$

Now for F_2 :

$$\sin \phi = \frac{F_{2x}}{F_2} ; \quad F_{2x} = 200 \, \text{N} \, (\sin 30^o) = 200 \, \text{N} \times 0.5 = 100 \, \text{N}$$

$$\cos \phi = \frac{F_{2y}}{F_2} ; \quad F_{2y} = 200 \, \text{N} (\cos 30^o) = 200 \, \text{N} \times 0.866 = 173 \, \text{N}$$

Watch Out!

Notice that the angle θ in the diagram is taken with respect to the y axis while ϕ was taken with respect to the x axis. When you are first learning how to resolve vectors into components it is important that you understand the details of what you are doing. Later, when you wish to improve your speed, you can develop some faster approaches.

Step 3. By looking at the diagram, determine how to add the existing x and y components to form the x and y components of the Resultant (R).

From the diagram we see that both F_{1x} and F_{2x} are in the direction of the $-x$ axis. Therefore for the x component of the Resultant we will write:

$$R_x = -F_{1x} - F_{2x} = -76.6 \, \text{N} - 100 \, \text{N} = -177 \, \text{N}$$

Now look back at the diagram and see how the y components of F_1 and F_2 are pulling against each other. F_{1y} is pulling in the $+y$ direction, while a stronger force, F_{2y} is pulling in the $-y$ direction. Therefore we write:

$$R_y = F_{1y} - F_{2y} = 64.3 \, \text{N} - 173 \, \text{N} = -109 \, \text{N}$$

Step 4. Draw another diagram that shows exactly where R_x and R_y are on the coordinate system.

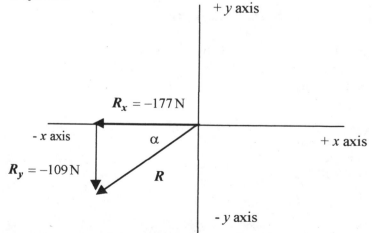

Step 5. To find the magnitude of the Resultant, use the Pythagorean Theorem:

$$R^2 = (-177)^2 + (-109)^2 \quad ; \quad R = 208\,\text{N}$$

Step 6. To find the direction of the Resultant, use any one of the trigonometric relationships:

$$\sin\alpha = \frac{109}{208} \quad ; \quad \alpha = \sin^{-1}\frac{109}{208} \quad ; \quad \alpha = 31.6^o$$

$$\tan\alpha = \frac{109}{177} \quad ; \quad \alpha = \tan^{-1}\frac{109}{177} \quad ; \quad \alpha = 31.6^o$$

To do this calculation on your calculator, you need to hit the inv or 2nd F key and then the sin or tan key.

Depending on your textbook or instructor, you may be asked to specify the angle of the Resultant in one of several possible forms. A diagram with an angle clearly marked as we have above may be acceptable. Another common form is to designate the angle with respect to the $+x$ axis. Going counterclockwise from the $+x$ axis, our angle is $180^o + 31.6^o = 211.6^o$. Some textbooks may refer to this angle as 31.6^o south of west.

The final answer is $R = 208\,\text{N}\angle 211.6^o$.

Let's consider the magnitude of this vector. Suppose we had gotten an answer of $310\,\text{N}$. Could this possibly be correct? The two forces that we are adding are $100\,\text{N}$ and $200\,\text{N}$. Even if these forces were pulling in the same direction they would only total $300\,\text{N}$. **Insight** Since the forces are pulling against each other somewhat, we expect a lower magnitude for the Resultant.

Now let's look at a weird and paranormal property of your calculator which will no doubt be picked up on by some of those in your class who like nerd-type things. Just for kicks, let's go back and re-solve for the x and y components of the Resultant on the last problem, but this time let's use the angle 211.6^o. Remember that's the angle to the Resultant from the $+x$ axis. To keep it simple, just use the magnitude of the hypotenuse of the Resultant triangle, 208 N and that angle in the following formulas:

$$R_x = 208\,\text{N}\cos(211.6^o) = 208\,\text{N}\times(-.852) = -177\,\text{N}$$

$$R_x = 208\,\text{N}\sin(211.6^o) = 208\,\text{N}\times(-.524) = -109\,\text{N}$$

When you took the $\cos(211.6^o)$ did you get -0.852 ? Your calculator knows that the x component of a vector with this angle from the $+x$ axis will be in the negative x direction. Also it knows that the $\sin(211.6^o)$ means that the y component is also negative. If you're used to thinking of sines and cosines in terms of giving you the side of a right triangle then it won't make much sense to take the sine of 211.6^o or to get a negative number for a side of a triangle. But fear not, this is not some advanced form of trigonometry. The calculator has just simply been programmed to give you the direction of a vector component if the angle from the $+x$ axis is specified. This explanation is included just to let you know what's going on if you see your instructor doing problems this way. If you're struggling with trig at all, we suggest going slower and drawing out the triangles. The authors are not confused at all about trig, and we still do the problems this way!

Remember, there are lots of cute ways to solve physics problems, but we think it is best to use the ways that you **understand.**

There are a great deal of problems in physics that require you to add or subtract vectors. Practice the procedure above on problems in the text. This is probably the most important type of problem that you need to know how to solve in the early part of the course. There are several variations of how vector addition problems are stated in words. For example, suppose that a car travels 10 miles north and then 15 miles 40° east of north. Both figures below are equivalent, but the diagram on the right may make it easier to add the vectors using the same procedure as outlined above.

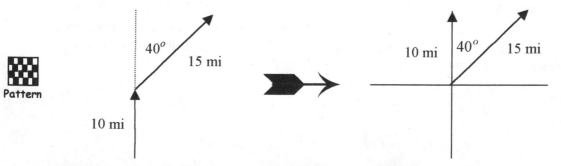

Consider the 15 mi vector above at an angle 40° east of north. The y component may be found in two ways:

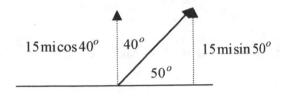

Verify on your calculator that $\cos 40^o = \sin 50^o = 0.766$. It is good to keep in mind that the components of a vector can be constructed from either triangle shown above. Use whichever triangle is more convenient.

2

MOTION IN ONE DIMENSION

Definitions and Terms

- Distance vs. displacement: Distance is a scalar quantity. If a jogger runs 2 miles north and then 3 miles south, the total distance travelled is 5 miles. The displacement is a vector pointing from the initial starting point to the final point. In this case the displacement is a vector with a magnitude of 1 mile pointed south.
- Speed vs. velocity: Speed is a scalar quantity. If a car is driven 100 miles in 2 hours, then the average speed of the car is 50 mi/hr. Velocity is a vector quantity, a number plus direction: 50 mi/hr at an angle of 10° west of north would be a velocity.
- Instantaneous quantities vs. average quantities: In the above example, we said that the average speed of the car was 50 mi/hr. This is the average for the whole 2-hour trip. Suppose that the car was driven at 60 mi/hr for the whole trip with the exception of a stop for gas along the way. How long was the gas stop in that case? Here's a chance to use that units conversion business that we learned in the last chapter. The "cruising" part of the trip was $100 \ mi \times \dfrac{hr}{60 \ mi} = 1.67 \ hours$. So the gas stop must have been the difference between 1.67 hours and 2 hours or 0.23 hours. Therefore, at any instant, the speed of the car may have been either 60 mi/hr or 0 mi/hr, but the average velocity for the whole trip is 50 mi/hr.
- The following 4 kinematic equations of motion with constant acceleration should be memorized:

Remember

$$v = v_0 + at \tag{2-1}$$

$$x = v_0 t + (1/2)at^2 \tag{2-2}$$

$$x = (1/2)(v_0 + v)t \tag{2-3}$$

$$v^2 = v_0^2 + 2ax \tag{2-4}$$

• The difference between a final value and initial value is usually referred to as "delta". Therefore, delta v = final velocity − initial velocity = $v - v_0$.

2-1 A train accelerates from rest to a speed of 30 meters per second in 12 seconds with a constant acceleration. What is the distance travelled by the train in the 12 seconds?

Solution: Well, let's see what we've got in terms of things that we know: Reading through the problem again, "from rest"....aha! that means $v_0 = 0$, the initial velocity is zero. We also have $v = 30$ m/s and $t = 12$ s. The first time you do these types of problems it is best just to write down the 4 equations of kinematics and all the quantities that are known and try to pick the right equation, so on a sheet of paper we would write:

Equations	Things we know
2-1. $v = v_0 + at$	$v_0 = 0$
2-2. $x = v_0 t + (1/2)at^2$	$v = 30$ m/s
2-3. $x = (1/2)(v_0 + v)t$	$t = 12$ s
2-4. $v^2 = v_0^2 + 2ax$	$x = ?$

OK, let's go through this. Equation 2-1 won't do us any good because there's no x in it. Equation 2-2 has an x, v_0 (which we know), t (which we know) –oh – but it also has an a in it. We have no idea what a is. All we know is that it is a constant. So we can't use equation 2-2 to solve for x because there would be two unknowns in the equation. I suppose, maybe, we could find another equation that would let us solve for a and then substitute that in, but let's look further in the list of equations to see if we can find just one equation that will give us the answer. Equation 2-3 looks like a winner, doesn't it? It has the variable we are looking for, x, expressed in terms of things that we know, v's and t.

$$x = (1/2)(30 \text{ m/s}) \times 12 \text{ s} = 180 \text{ m}$$

2-2 Continuing the problem above, suppose that once the train reaches 30 m/sec, the operator of the train applies the brakes and produces a constant deceleration of 3 m/s^2. How far does the train go before it stops?

Equations	Things we know
2-1. $v = v_0 + at$	$v_0 = 30$ m/s
2-2. $x = v_0 t + (1/2)at^2$	$v = 0$ m/s
2-3. $x = (1/2)(v_0 + v)t$	$a = -3$ m/s^2
2-4. $v^2 = v_0^2 + 2ax$	$x = ?$

Solution: Equation 2-4 is the one that contains everything we know, except for what we're looking for. Notice this time that the initial velocity, v_0, is 30 m/s and the final velocity is 0 m/s. The acceleration is negative: that's how to handle deceleration.

$$0^2 = 30^2 + 2(-3)x \qquad \text{or} \qquad (0 \text{ m/s})^2 = (30 \text{ m/s})^2 + 2(-3 \text{ m/s}^2)x$$

$$-900 = -6x \implies x = 150 \text{ meters}$$

A comment about the units: Textbooks usually like to write the units down in every equation. It helps to clarify where each quantity came from. On a test you may not want to write the units out for each equation for reasons of time, but you should at least write them out for the first equation of the problem.

2-3 Two trains are travelling along a straight track, one behind the other. The lead train is travelling at 15 m/s and continues to travel at constant velocity. The second train, approaching from the rear, is travelling at 30 m/s. When the second train is 200 meters behind the first, the operator applies the brakes producing a constant deceleration of 0.25 m/s^2. At what time after the operator applies the brakes do the trains collide?

Solution: Now if you were to look back at the four equations of motion and start thinking about which one to use, you would hopefully at some point come to the realization that the problem isn't that simple. This problem can't be solved by just plugging numbers into a single equation. There are two objects moving here. Therefore we present a more organized approach to solving these types of problems, which is very effective on the more difficult problems. The key thought process in the beginning is that we are **not** going to try to plan out in our minds exactly how we are going to get the answer to the problem. For many people this only causes more confusion. This is a very important point though, not only in solving physics problems, but also in solving scientific puzzles in general. Many times it is more important that you know how to recognize a way of redefining the problem in a more organized way than immediately "seeing" the way to the answer.

Step 1: Understand the details of the problem. We must make sure in the beginning that we clearly understand what situation the problem is describing and what the numbers given in the problem really mean. The path to the answer is in the back of our minds at this stage. For now we are focusing on understanding the question. Do not underestimate the amount of time you should spend just trying to understand the question. Since most people prefer to visualize, let's draw a diagram:

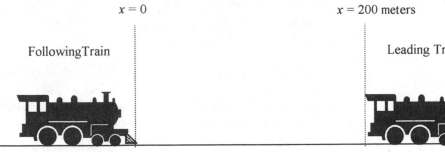

$x = 0$ $x = 200$ meters

FollowingTrain   Leading Train

$v_{0\,follow} = 30$ m/s $v_{lead} = 15$ m/s

$a_{follow} = -0.25$ m/s^2 $a_{lead} = 0$

Again, we are going to write down the four equations of motion. Now, go back and read the problem again. Isn't the whole question of x a little confusing? We know that the trains **start** 200 meters apart, but exactly where they collide isn't immediately obvious. One thing that you might be able to "see" at this point is that regardless of how far the following train travels before it collides with the lead train, it must travel 200 meters farther than the lead train.

> **Equations**
>
> 2-1. $v = v_0 + at$
>
> 2-2. $x = v_0 t + (1/2)at^2$
>
> 2-3. $x = (1/2)(v_0 + v)t$
>
> 2-4. $v^2 = v_0^2 + 2ax$

Now at this point you may be thinking, who cares about x!
The problem is asking **when** the trains collide, not **where**! Understand this about step 1 of this process: remember we are trying to clearly understand the question first and what all of the numbers we have been given in the problem **really mean**. If you go too fast here and start plugging 200 meters in for x into one of the four equations along with other values of v and a from the problem you're headed for trouble.

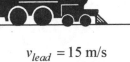

Watch Out!

Step 2: Using the four equations of motion in one dimension, and the numbers given in the problem, construct equations for position, velocity, and acceleration for each of the moving objects in the problem. In other words get $x =$ something, $v =$ something, and $a =$ something for each train. The key here is that since we don't immediately see how to get the answer, we're just going to start generating equations and then see where this leads. Let's take $t = 0$ to mean the time when the brakes are first applied. So the equations will be:

$x_{following} = (30 \text{ m/s})t + (1/2)(-0.25 \text{ m/s}^2)t^2$ from equation 2-2

$v_{following} = 30 \text{ m/s} + (-0.25 \text{ m/s})t$ from equation 2-1

$a_{following} = -0.25 \text{ m/s}^2$ from the question

$$x_{leading} = 200 \text{ m} + (15 \text{ m/s})t$$ This came from equation 2-2 ($a = 0$) and we also added 200 m because the lead train started at x = 200 m.

$$v_{leading} = 15 \text{ m/s}$$

$$a_{leading} = 0$$

The trickiest part of this is that we didn't plug in 200 meters for x anywhere. The variable x is interpreted as the distance each train travels from the $x = 0$ point between the time the brakes are applied ($t = 0$) and the time of the collision. So actually $x_{following} = x_{leading}$.

Step 3: Now we think about how to find t, and solve for t. If we know that $x_{following} = x_{leading}$, then we can use that to form an equation where t is the only variable:

$$(30 \text{ m/s})t + (1/2)(-0.25 \text{ m/s}^2)t^2 = 200 \text{ m} + (15 \text{ m/s})t$$

Now it's just algebra to get the value of t.

$$-0.125t^2 + 30t - 15t = 200$$

$$-0.125t^2 + 15t - 200 = 0$$

$$t = \frac{-15 \pm \sqrt{15^2 - 4(-0.125)(-200)}}{2(-0.125)}$$

$$t = \frac{-15 \pm \sqrt{225 - 100}}{-0.25} = \frac{-15 \pm 11.2}{-0.25}$$

$$t = \frac{-15 + 11.2}{-0.25} = 15.2 \text{ seconds}$$

OR $$t = \frac{-15 - 11.2}{-0.25} = 59.8 \text{ seconds}$$

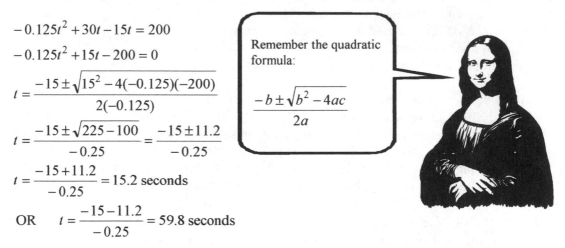

Remember the quadratic formula:

$$\frac{-b \pm \sqrt{b^2 - 4ac}}{2a}$$

Even though the mathematics gives us two solutions to the quadratic formula, the physics would dictate that the first time would be the time of the collision. The answer then is that the trains collide 15.2 seconds after the brakes are applied.

2-4 Galileo drops a cannon ball weighing 10 lbs and a small rock weighing 2 lbs off the leaning tower of Pisa. He releases both from a height of 120 meters at the same time. Assuming there is no air resistance, when does each object hit the ground?

Solution: First, both objects hit the ground at the same time. If a feather and a cannon ball were dropped, the feather would take a lot longer to reach the ground, but not because the force of gravity on it is any different. The difference would be the air resistance on the way down. The way the force of gravity between two objects works is that both objects attract each other and the force of the attraction depends on how massive the objects are. So for example on the moon, both objects would take longer to fall. The masses of the rock and the cannon ball are so much less than the mass of the earth that the force of gravity on them is for all practical purposes the same and they have the same acceleration. (Also on the moon there is no air, so the feather would hit the ground at the same time too.) The value of that acceleration is $a_{gravity} = g = 9.8 \text{ m/s}^2$.

Also neither of these objects is changing its distance from the center of the earth significantly so that the acceleration remains constant. At the distance of orbiting satellites from the earth, the acceleration would be less, but small variations in height near the surface of the earth aren't significant. Now let's get to the problem:

Equations	Version 1 Down is positive Things we know	Version 2 Down is negative Things we know
2-1. $v = v_0 + at$	$v_0 = 0$ m/s	$v_0 = 0$ m/s
2-2. $x = v_0 t + (1/2)at^2$	$x = +150$ meters	$x = -150$ meters
2-3. $x = (1/2)(v_0 + v)t$	$a = +9.8 \text{ m/s}^2$	$a = -9.8 \text{ m/s}^2$
2-4. $v^2 = v_0^2 + 2ax$	$v = ?$	$v = ?$

Using equation 2-2,

Version 1
$$150\text{m} = 0 + (1/2)(9.8 \text{ m/s}^2)t^2$$
$$\sqrt{\frac{150}{4.9}} = t$$

Version 2
$$-150 \text{ m} = 0 + (1/2)(-9.8 \text{ m/s}^2)t^2$$

the minus signs cancel giving the same answer

$$t = 5.53 \text{ seconds}$$

2-5 From the height of 150 meters, suppose Galileo throws the rock straight up with an initial velocity of 10 m/s. (a) How long does it take the rock to return to Galileo's height above the ground (150 meters)? (b) How long does it take for the rock to hit the ground from the time it passes Galileo on its way back down?

Insight

Solution: (a) The motion of the rock is symmetric. The time that it takes the rock to get to its highest point and the time that it takes the rock to get back to its original height is the same. Also the velocity that the rock leaves with going up is the same as the velocity downward when the rock gets back to Galileo's height above the ground. The strategy, then, for part (a) will be to find the time that it takes for the ball to reach its maximum height (and the key there is to set $v = 0$, the final velocity equal to 0) and then multiply that answer by 2.

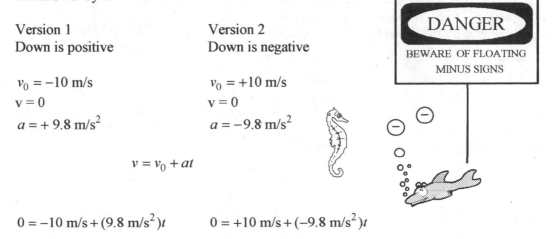

Version 1
Down is positive

Version 2
Down is negative

$v_0 = -10$ m/s

$v_0 = +10$ m/s

$v = 0$

$v = 0$

$a = +9.8$ m/s^2

$a = -9.8$ m/s^2

$$v = v_0 + at$$

$0 = -10 \text{ m/s} + (9.8 \text{ m/s}^2)t \qquad 0 = +10 \text{ m/s} + (-9.8 \text{ m/s}^2)t$

Now what's the deal with 2 versions and all these minus signs floating around? The reason for this is that textbooks and instructors differ. Sometimes they take the downward direction to be positive, sometimes negative. Some instructors are very rigid, they always label down as negative. Others take down as negative only on Tuesdays and Thursdays, except for a leap year, when down is negative on Mondays, Wednesdays, and Fridays. For those instructors that appear more flexible about directions, what they are really doing is taking the most convenient direction as positive (it's convenient to them because they tried the problem the other way and realized there were more minus signs floating around). Our view is this: It doesn't matter!

     The important thing to keep in mind for the above problem is that we are throwing the rock up with an initial velocity and gravity is pulling it down. Therefore we know that two terms in the equation must be opposite in sign. Notice that either version of how we pick the sign of the direction produces the same answer, which is 1.02 seconds. Remember that this is the time it took for the rock to reach its maximum height, so the answer to part (a) is twice that, or 2.04 seconds.

So now we are ready to go on to part (b) of the question, which is how long from this point does it take the rock to hit the ground.

	Version 1 Down is positive	Version 2 Down is negative
Equations	Things we know	Things we know
2-1. $v = v_0 + at$	$v_0 = 10$ m/s	$v_0 = -10$ m/s
2-2. $x = v_0 t + (1/2)at^2$	$x = +150$ meters	$x = -150$ meters
2-3. $x = (1/2)(v_0 + v)t$	$a = +9.8$ m/s^2	$a = -9.8$ m/s^2
2-4. $v^2 = v_0^2 + 2ax$	$v = ?$	$v = ?$

In this case, the initial velocity is in the same direction as the acceleration, so everything is either positive or negative depending on how we choose the sign of the direction. Using equation 2,

$$150 \text{ m} = (10 \text{ m/s})t + (1/2)(9.8 \text{ m/s}^2)t^2$$

$$0 = -150 + 10t + 4.9t^2$$

$$t = \frac{-10 \pm \sqrt{10^2 - 4(4.9)(-150)}}{2(4.9)}$$

$$t = \frac{-10 \pm 55.2}{9.8} = 4.61 \text{ seconds}$$

OK, let's review this problem. Galileo tossed a rock straight up with an initial velocity of 10 m/s from a height of 150 m above the ground. It took 1.02 seconds for the rock to get to its maximum height and stop. Then it took another 1.02 seconds to get back to Galileo's original height at which point it was moving downward with that same velocity of 10 m/s. From that point it took another 4.61 seconds to hit the ground. This last part makes sense because looking back at the previous problem we know that it took longer, 5.53 seconds, for the rock to fall the 150 meters when it had no initial velocity in the downward direction.

Now that I've figured out gravity, it's time to go build my telescope.

3

MOTION IN TWO DIMENSIONS

3-1 An aircraft carrier is crossing the Chesapeake Bay, moving perpendicular to the tide. The distance between the carrier and the shore is 5 miles and the tide flows with a speed of 5 mi/hr. The carrier has constant speed of 20 mi/hr. (a) When does the carrier reach the opposite side? (b) How far has the carrier been pushed by the tide by the time it reaches the opposite shore?

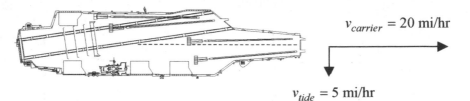

$v_{carrier} = 20$ mi/hr

$v_{tide} = 5$ mi/hr

Solution: (a) The idea here is that if the tide is moving at 5 mi/hr then the carrier is being carried along at this same speed, since it is propelling itself completely perpendicular to the tide. The first important thing to understand about this situation is that as long as the carrier is moving perpendicular to the tide (that means at a 90° angle to the tide), it will always take the same amount of time to travel the 5 miles, regardless of the speed of the tide. *The speed in the x direction is completely independent of the speed in the y direction.* So the time to travel the 5 miles is just:

$$\frac{5 \text{ miles}}{} \times \frac{1 \text{ hour}}{20 \text{ miles}} = 0.25 \text{ hour} \quad \text{or} \quad 15 \text{ minutes}$$

$35m \times \dfrac{20m}{\text{hs}} =$

(b) Now that we know how much time is involved we can do a similar calculation to see how far the carrier is pushed by the tide:

$$\frac{5 \text{ miles}}{\text{hour}} \times \frac{0.25 \text{ hour}}{} = 1.25 \text{ miles}$$

If the tide were moving faster, the carrier would be pushed further in a direction perpendicular to its motion; however, the time to travel the 5 miles would remain the same.

3-2 A police officer fires his handgun at a specific height and parallel to the ground and at the same time, drops a bullet from the same height. (a) Which bullet hits the ground first? (b) If the bullet fired from the gun travels 1000 feet per second, and is fired from a height of 5 feet, how far does the bullet travel horizontally before it hits the ground?

Solution: (a) Both bullets hit the ground at the same time. Remember that as long as the gun is fired exactly perpendicular to the acceleration caused by gravity then it doesn't matter how fast the bullet travels. The vertical motion of the bullet cannot be changed by the horizontal movement of the bullet. (b) To find out how far away the bullet is when it hits the ground, first we need to find the time it takes for the bullet to fall. For this calculation, use one of those equations of motion from chapter 2.

Equations	Things we know
2-1. $v = v_0 + at$	$v_{0y} = 0$
2-2. $x = v_0 t + (1/2)at^2$	$y = 5$ feet
2-3. $x = (1/2)(v_0 + v)t$	$a_y = 9.8$ m/s^2
2-4. $v^2 = v_0^2 + 2ax$	

There is more than one subscript being used now, so that we can *distinguish between the equations of motion in the x and y directions*. Also, remember that in the equations from chapter 2 we can use them for the *x* direction as well as the *y* direction. The answer to how long it takes the bullet to fall can be found using the second equation above:

$$y = v_{0y}t + (1/2)a_y t^2 \implies 5 \text{ feet} = 0 + (1/2)(9.8 \text{ m/s}^2)t^2$$

Watch Out!

Whoops, that's not going to work yet. We've got feet on one side of the equation and meters on the other. Let's convert feet to meters first.

$$\frac{5 \text{ feet}}{} \times \frac{.3048 \text{ meter}}{1 \text{ foot}} = 1.52 \text{ meters}$$

$$1.52 \text{ m} = (4.9 \text{ m/s}^2)t^2 \implies t = \sqrt{\frac{1.52}{4.9}} = 0.56 \text{ second}$$

Now we know that the bullet will be in the air travelling horizontally at 1000 feet/sec for 0.56 seconds at which point it will strike the ground. The answer to part (b) of the question then is:

$$x = \frac{1000 \text{ feet}}{\text{sec}} \times \frac{0.56 \text{ sec}}{} = 560 \text{ feet} .$$

In this problem remember that we assumed there was no air resistance and that the bullet moves at constant velocity.

3-3 A sailboat is moving in the water at a speed of 5 mi/hr. A sailor climbs the mast on the boat to tighten a screw at the top of the mast. When near the top of the mast, at a height of 40 feet above the deck, he accidentally drops his screwdriver. The sailor imparts no horizontal velocity to the screwdriver, it simply falls vertically. Where on the deck below him does the screwdriver land?

Solution: The screwdriver lands on the deck directly below the point where it was dropped. There is no calculation needed here. The boat, the sailor on the mast, and the screwdriver all start with a speed of 5 mi/hr in the horizontal direction, and they all end with that same horizontal speed. They are all moving together and they all have the same speed relative to the water.

3-4 An airliner is approaching a runway for a landing. The pilot of the airliner measures her horizontal airspeed to be 400 miles per hour. She hears from ground control that she is moving through a head wind of 20 mi/hr. The altimeter indicates that the plane is dropping at a speed of 10 feet per second. If the plane is at a height of 3,000 feet and the start of the runway is 28 miles away (horizontally), does the pilot need to make any adjustments to her descent or will she land the plane at the start of the runway?

Solution:

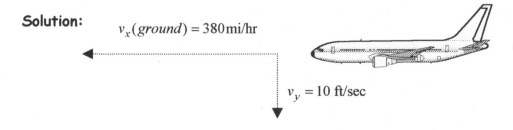

$$v_x(ground) = 380 \text{mi/hr}$$

$$v_y = 10 \text{ ft/sec}$$

Pattern

The strategy here is to calculate the time it takes for the plane to descend to the ground and then see how far horizontally the plane travels in that time. This is a pattern that comes up quite often in projectile motion problems. We saw it earlier in the question about the falling bullet.

The first thing to straighten out in this problem is that the speed of the plane relative to the air may be 400 mi/hr, but since there is a 20 mi/hr head wind, the speed of the plane relative to the ground is only 380 mi/hr. So for the purposes of calculating the approach to the runway, we use the 380 mi/hr. Now first let's deal with the y direction and find out how much time the pilot has before the plane hits the ground:

$$\frac{1 \text{ second}}{10 \text{ feet}} \times \frac{3000 \text{ feet}}{} = 300 \text{ seconds} \times \frac{1 \text{ min}}{60 \text{ sec}} = 5 \text{ minutes}$$

We are not invoking any formulas here. We are simply using the rate of descent and the height above the ground to find out how much time is involved. (Since the decent is at constant velocity this part of the problem is actually easier than the bullet problem, because the bullet was accelerating in the vertical direction under the force of gravity.) OK, the pilot has 5 minutes of flying time at her current rate of descent. Now let's see how far she can go in that time.

$$5 \text{ minutes} \times \frac{1 \text{ hr}}{60 \text{ minutes}} = .0833 \text{ hr} \times \frac{380 \text{ mi}}{1 \text{ hr}} = 31.7 \text{ miles}$$

Now we can answer the question. Since the runway is 28 miles away and at the current rate of descent the plane will travel 31.7 miles before touching the ground, the pilot needs to slow down in order to land. Notice that every problem can't be solved by simply memorizing those 4 equations of motion given in chapter 2 and plugging numbers in. You also need to be able to think in terms of dimensional analysis and be able to convert the units of the quantities given in the problem.

3-5 One of the Apollo astronauts, Alan Shepard, was the first man to hit a golf ball on the moon. Suppose that he places the ball on the surface of the moon and hits it with his golf club. The ball is launched with a velocity of 50 meters/second at an angle of 35° above the horizontal. Given that the acceleration due to gravity on the moon is $1/6^{th}$ of the acceleration on the earth and neglecting air resistance (which is easy to neglect since there is almost no air on the moon), (a) How far does the golf ball travel before it hits the ground? (b) What is the time of flight? (c) With what velocity and at what angle does the ball hit the ground?

Solution: The first thing to realize about this problem is that the horizontal component of velocity will always be the same. Start thinking about how you are going to separate out the initial horizontal velocity and the initial vertical velocity. Part (c) is the easiest part if you know the secret. Remember the problem in chapter 2 (2-4) where Galileo tossed a rock straight up and the question was asked how long is the rock in the air before it comes back down to the point where it was thrown? There we said that the time for the rock to rise to its maximum height and stop was the same as the time for the rock to fall from its maximum height back to the point where it was released. Also we said that the magnitude of the vertical velocity of the rock when it came back down to the point where it was first thrown upward was the same as the initial upward velocity. The same is true in this problem. Just because there is a horizontal component of velocity doesn't mean that anything happening in the vertical direction is any different. Remember x and y components of velocity are completely independent of one another. So, if we know that the horizontal component of velocity never changes, and we know that by the symmetry of the problem the vertical component of velocity will have the same magnitude when the golf ball lands back on the surface of the moon, then the answer to part (c) of this question is easy. The answer to part (c) is that the ball hits the ground with the same speed that it left the ground and at the same angle of 35° with respect to the horizontal. OK, now draw a picture of the problem:

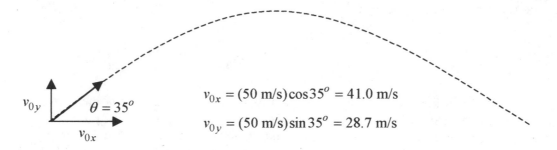

$$v_{0x} = (50 \text{ m/s})\cos 35^o = 41.0 \text{ m/s}$$

$$v_{0y} = (50 \text{ m/s})\sin 35^o = 28.7 \text{ m/s}$$

Now, since this problem is a little more complicated than ones we've seen in the past, let's use a strategy of writing down x = something, v = something, and a = something for both the x and y directions:

x direction	y direction	Equations
$x = v_x t$	$y = v_{y0}t + (1/2)g_{moon}t^2$	2-1. $v = v_0 + at$
v_x = constant	$v_y = v_{y0} + g_{moon}t$	2-2. $x = v_0 t + (1/2)at^2$
$a_x = 0$	$a_y = g_{moon} = (1/6)9.8 \text{ m/s}^2$	2-3. $x = (1/2)(v_0 + v)t$
		2-4. $v^2 = v_0^2 + 2ax$

Why didn't we use equation 2-4 to describe the motion for the y component of velocity? The answer is that the unknown that it contains is x (or y), and we are interested in t as the unknown in order to answer part (b) of the question. In looking at these equations, do you see that $x = v_x t$ is how we arrived at the answer to part (a)? But first, we need t, the time of flight. We can get the time of flight by using the second equation for the y direction.

We will calculate how long it takes the golf ball to reach its maximum height and then double that time.

$$v_y = 0 = (28.7 \text{ m/s})t + (1/2)(\frac{-9.8 \text{ m/s}^2}{6})t^2 \qquad \textit{time to reach maximum height}$$

$$0 = 28.7t - .817t^2$$

$$0 = 28.7 - .817t \implies t = \frac{28.7}{.817} = 35.1 \text{ seconds}$$

The total time of flight is then $2 \times 35.1 = 70.25$ seconds . This is the answer to part (b) of the question.

The horizontal distance travelled in this time is:

$$x = (41.0 \text{ m/s})70.3 \text{ sec} = 2882 \text{ meters} \implies 1.8 \text{ miles}$$

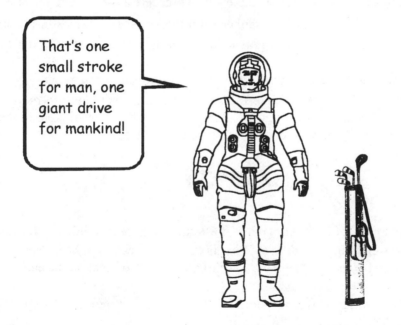

That's one small stroke for man, one giant drive for mankind!

4

FORCES

Things to Remember

- $F = ma$. Also memorize the mks units Newtons $= kg \times meters/second^2$.
- An object **weighs** less on the moon, but it has the same **mass**.
- For objects close to the surface of the earth, the acceleration of gravity is equal to "little g" or 9.8 m/s^2. For objects farther from the surface of the earth (10 km and above, roughly) use the universal gravitational constant, big G, to find the force of gravity.
- Friction always has the general form $f_k = \mu_k F_N$ for kinetic friction and $f_s = \mu_s F_N$ for static friction. The force of static friction, or the force that needs to be overcome to start a body moving, is always greater than the force of kinetic friction, or the force needed to keep a body moving once it has started in motion.

GRAVITY

4-1 A car has a mass of 2200 kilograms. (a) What is the car's weight in Newtons? (b) What is the car's weight in pounds? (c) What is the car's mass, weight in Newtons, and weight in pounds on the moon, where the force of gravity is $1/6^{th}$ that on the earth?

Solution: The first thing to realize here is that weight is a force. (a) Using $F = ma$, $2200 \text{ kg} \times 9.8 \text{ m/s}^2 = 21{,}560 \text{ N}$. (b) Pounds are also a force. The conversion factor from pounds to Newtons is 1 lb $= 4.448$ N, so $21{,}560 \text{ N} \times \dfrac{1 \text{ lb}}{4.448 \text{ N}} = 4847 \text{ lbs}$. (c) On the moon, the mass of the car is the same. Mass is an inherent property of matter. The weight in Newtons will be $\dfrac{21{,}560}{6}$ or 3593 N. The weight in lbs is $\dfrac{4847}{6}$ or 808 lbs.

4-2 Suppose that you step onto an elevator. The doors shut and there are no windows. You notice that next to you is Albert Einstein. He is standing on a scale that reads his weight, 145 lbs Einstein speaks: "The scale reads my weight due to the force of gravity, BUT! – Since there are no windows on the elevator it would be impossible for us to tell the difference between standing on a scale in the presence of the earth's gravitational field, OR being out in space with no gravitational force on us at all, but accelerating. You see, accelerations produce forces just like gravitational fields do, and inside this elevator it would be impossible to do any experiment to tell whether the force that I read on this scale below me is due to gravity or acceleration. I call this the equivalence principle. It is part of my theory of general relativity."

Insight

Einstein then hits the button for the 25th floor. The elevator begins to move upward. The scale now reads 165 lbs "See, the acceleration of the elevator is now causing a greater force to be read by the scale. But soon the elevator will reach a constant velocity." As the elevator reaches a constant velocity the reading on the scale returns to 145 lbs. As the elevator slows, the scale reads 125 lbs. The elevator then stops at the 25th floor and the doors open. As Einstein exits he says, "As an exercise you should calculate the curvature of space-time produced by the gravitational field of the earth. In addition, you can also calculate the acceleration of the elevator as we moved upward earlier." He then vanishes into thin air.

Solution: The first part of Einstein's question is better left to the weirdos taking general relativity. As for the second part, we know that Einstein's apparent weight increased by 20 lbs when the elevator accelerated upward. Let's convert this to Newtons: $20\,\text{lbs}\dfrac{4.448\,\text{N}}{\text{lb}} = 89\,\text{N}$. We know the force on Einstein produced by the elevator. What we need now is his mass. Then we can use $F = ma$ to get the acceleration of the elevator. To get Einstein's mass we can use the fact that we know his weight at rest. Let's get that in Newtons: $145\,\text{lbs} \times \dfrac{4.448\,\text{N}}{\text{lb}} = 645\,\text{N}$. Now to find his mass:

$$645\,\text{N} = m(9.8\,\text{m/s}^2) \quad \Rightarrow \quad m = 645/9.8 = 65.8\,\text{kg}$$

So the acceleration of the elevator (which is the same as Einstein's acceleration since they are not moving relative to each other) is the force on Einstein divided by his mass, or $89\,\text{N}/65.8\,\text{kg} = 1.35\,\text{m/s}^2$.

4-3 Consider four cases: a mass on the surface of the earth, and at a height of 1 meter, 10 km, and 1000 km. In which of these cases does the acceleration due to gravity on an object change significantly?

Solution: The first thing to realize about this problem is that it is a "big G" problem. You need to remember the difference between little g and big G and when you use them.

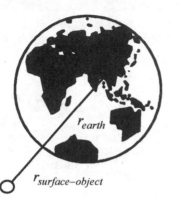

Little g is 9.8 m/s^2. This is the acceleration due to gravity **near the surface of the earth.** For satellites orbiting the earth or for gravitational forces between the earth and moon, we need to use the general equation for the gravitational force, which is:

$$F_{gravity} = G\frac{m_1 m_2}{r^2}$$

Remember

where m_1 and m_2 are the two masses involved, r is the distance between the **center (remember that – the center)** of the two masses, and big G is the universal gravitational constant. The force effectively acts along the line connecting the center of the two masses. *Universal* means that big G can be used for anything. It works for the sun, the moon, the earth, Mars, a spec of dust – anything. For the earth we write:

$$F_{gravity} = \text{Weight(in Newtons)} = G\frac{m_{earth} m_2}{r^2} = m_2 a = m_2 g$$

From this can you see that little g is: $g = G\frac{m_{earth}}{r^2}$. The value of r^2 is the radius of the earth plus the distance of the object from the surface of the earth. The radius of the earth is 6.38×10^6 meters. Therefore, the value of r for the four cases would be 6.38×10^6 m, 6.38×10^6 m $+1$ m, 6.38×10^6 m $+1\times10^4$ m or 6.39×10^6 m, and 6.38×10^6 m $+1\times10^6$. There is no practical difference between the first and second case - adding one meter to over 6 million doesn't really matter. In the second case, we are just barely seeing a significant change in the value of r. By the time we get to the last case, 1,000 km or 1,000,000 meters, we now see a significant change in r. The reason for going through all this is to show that little g is a very practical approximation for the acceleration due to gravity at distances close to the earth – say up to 10 km or so. For farther distances it is necessary to use the big G equation.

4-4 One possible orientation of the sun, earth, and moon is shown in the following diagram. Find the net gravitational force on the moon given the following:

$m_{sun} = 2 \times 10^{30} \, \text{kg}$, $m_{earth} = 6 \times 10^{24} \, \text{kg}$, $m_{moon} = 7.35 \times 10^{22} \, \text{kg}$, $r_{sun-moon} = 1.5 \times 10^{11} \, \text{m}$, $r_{earth-moon} = 3.85 \times 10^{8} \, \text{m}$

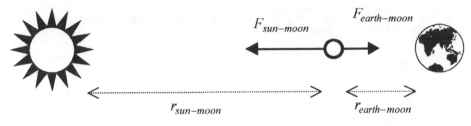

Solution: This is a big G problem. The forces of gravity can be found by:

$$F_{earth-moon} = G \frac{m_{earth} m_{moon}}{r_{earth-moon}^2} = 6.67 \times 10^{-11} \, \text{N} \, \text{m}^2/\text{kg}^2 \frac{6 \times 10^{24} \, \text{kg} \times 7.35 \times 10^{22} \, \text{kg}}{(3.85 \times 10^{8} \text{m})^2} = 1.98 \times 10^{20} \, \text{N}$$

$$F_{sun-moon} = 6.67 \times 10^{-11} \, \text{N} \, \text{m}^2/\text{kg}^2 \frac{2 \times 10^{30} \, \text{kg} \times 7.35 \times 10^{22} \text{kg}}{(1.5 \times 10^{11} \text{m})^2} = 4.36 \times 10^{20} \, \text{N}$$

So the net force on the moon in this orientation is actually directed at the sun and has a magnitude of about $2.38 \times 10^{22} \, \text{N}$. Interesting isn't it? Does this mean that the moon should fall into the sun? No. It is true that if the earth and the moon were both at rest relative to the sun, then the sun would pull the moon away from the earth. However both the moon and the earth have orbital velocity around the sun. The study of how objects with orbital velocity effectively balance out gravitational forces is contained in the next chapter on centripetal force.

4-5 A car accelerates from rest and reaches a speed of 60 mi/hr in 6.0 seconds. The mass of the car is 1600 kg. What is the magnitude of the force that acts on the car?

Solution: We know that $F = ma$. We have the mass. The acceleration can be found by using one of the linear equations of motion from chapter 2. So the plan is to solve for acceleration and then substitute that into $F = ma$ to get the force.

Equations	Things we know	
2-1. $v = v_0 + at$	$v_0 = 0$	
2-2. $x = v_0 t + (1/2)at^2$	$v = 60$ mi/hr	
2-3. $x = (1/2)(v_0 + v)t$	$t = 6.0$ sec	$m = 1600$kg
2-4. $v^2 = v_0^2 + 2ax$	$a = ?$	

First, convert that final velocity to meters per second:

$$\frac{60\,\text{mi}}{\text{hr}} \times \frac{1\,\text{hr}}{60\,\text{min}} \times \frac{1\,\text{min}}{60\,\text{sec}} \times \frac{5280\,\text{feet}}{1\,\text{mi}} \times \frac{1\,\text{meter}}{3.281\,\text{feet}} = 26.8\,\text{meters/sec} = 26.8\,\text{m/s}$$

Using equation 2-1

$$26.8\,\text{m/sec} = 0 + a(6.0\,\text{sec}) \implies a = 4.47\,\text{m/sec}^2$$

$$F = ma = 1600\,\text{kg} \times 4.47\,\text{m/sec}^2 = 7152\,\text{kg}\cdot\text{m/sec}^2 = 7152\,\text{Newtons}$$

4-6 Testing the Normal Force: A person places their hand on top of a table and under a 10 pound block. What is the force on the top of their hand and what is the force on the bottom of their hand?

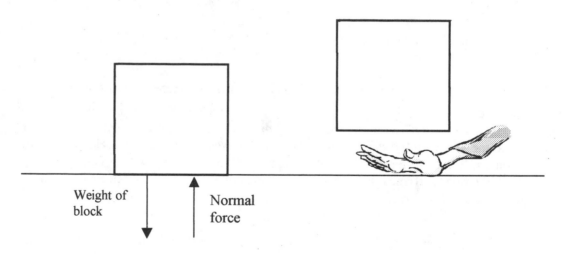

Weight of block

Normal force

Solution: The answer is that there is an equal force on the bottom of their hand as on the top. The force on the top of their hand (10 lbs) is the weight of the block. Pounds is a unit of force, by the way. The force on the bottom of the hand is called the normal force. Think of this as the force of the table pushing on the hand which in turn supports the weight of the block. If the block were to be resting directly on the surface of the table there would still be a normal force directed upward supporting the weight of the block. The hand in this situation acts like a force meter, able to feel forces in both directions. If you ever have trouble understanding normal force, just put your hand under a heavy book and see if you feel pressure on both sides of your hand.

Insight

Another way of thinking about this problem is that net forces (or unbalanced forces) produce accelerations. The block isn't moving when it is resting on the table, and we know that the force of gravity is acting on it in the downward direction. Since the block is not accelerating downward, there must be a force equal to the force of gravity acting in the opposite direction – this is the normal force.

4-7 A man wishes to slide a box along the ground by pushing it. He pushes down and horizontally on the box, applying a force of 282 Newtons at an angle of $20°$ below the horizontal. The box has a mass of 100 kg. The coefficient of static friction between the box and the floor is 0.25. Is he able to start the box moving?

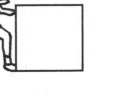

Solution: First, draw a vector diagram of all the forces acting on the box (some textbooks like to call these "free body diagrams"). Draw all the forces acting in both the x and y directions. Make it a habit to draw these vector diagrams when you start a problem. **Pattern**

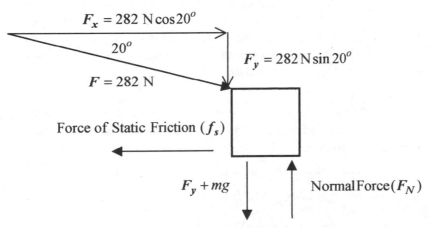

$F_x = 282 \text{ N} \cos 20°$

$20°$

$F_y = 282 \text{ N} \sin 20°$

$F = 282 \text{ N}$

Force of Static Friction (f_s)

$F_y + mg$

Normal Force (F_N)

In the x direction, there is the x component of the force of the man pushing, acting toward the right (or $+x$ direction). This is labeled $F_x = 282 \text{ N} \cos 20°$. This force needs to overcome the force of static friction, which acts in opposition to the pushing force. The force of static friction (f_s) is always equal to the horizontal pushing force that it opposes, up to the point where the block starts to move (if it moves). What we need to know is how high the force of static friction can get. How much pushing force can possibly be opposed by static friction, and are we above or below that force? The force of static friction always has the general form $f_s = \mu_s F_N$, where μ_s is the coefficient of static friction and F_N is the normal force. Now we need to calculate F_N. Up until now we have been thinking of F_N as the force that opposes the weight of an object, but look closely at this situation. The y component of the pushing force is pushing down on the box adding to the weight of the box and to the total downward force. The normal force therefore opposes both the weight of the box and the y component of the pushing force.

$$F_N = mg + 282 \text{ N} \sin 20^o = (100 \text{ kg})9.8 \text{ m/s}^2 + 282 \text{ N}(0.342) = 1076 \text{ Newtons}$$

OK, with this we can get the maximum value for the force of static friction:

$$f_s = \mu_s F_N = 0.25(1076 \text{ N}) = 269 \text{ N}$$

Now let's compare this to the horizontal pushing force on the box:

$$F_x = 282 \text{ N} \cos 20^o = 245 \text{ N} \quad \text{wrong}$$

244 N

It has been a while, so do you see where these numbers came from? The 282 N is the total pushing force given in the problem. This force was at an angle and we are dividing it up into an x component (which we just calculated as 245 N) and a y component, which effectively adds to the weight of the box. What we just found out, though, is that the x component of the pushing force is less than the maximum force of static friction. So the answer to the question is that the box doesn't move because the man is not pushing hard enough to overcome the force of static friction.

4-8 Fortunately, the man who is trying to slide the box has a wife who is taking physics. She is watching his unsuccessful attempt to push the box and has the physical intuition to see the solution to his problem.

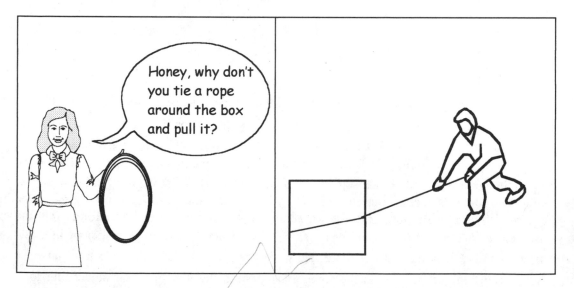

This time the man pulls the box with the same force of 282 N at a direction of 20° above the horizontal. Does the box start to move in this case?

Solution: Again, let's draw the vector diagram for the problem:

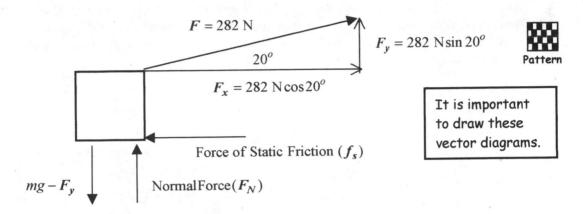

$F = 282\ \text{N}$

$F_y = 282\ \text{N}\sin 20^o$

20°

Pattern

$F_x = 282\ \text{N}\cos 20^o$

Force of Static Friction (f_s)

It is important
to draw these
vector diagrams.

$mg - F_y$

$\text{Normal Force}(F_N)$

The difference in this case is that the y component of the force used to move the box now subtracts from mg. The normal force then is recalculated as:

$$F_N = mg - 282\ \text{N}\sin 20^o = (100\ \text{kg})9.8\ \text{m/s}^2 - 282\ \text{N}(0.342) = 883\ \text{Newtons}$$

The maximum that the force of static friction can be is now:

$$f_s = \mu_s F_N = 0.25(883\ \text{N}) = 221\ \text{N}$$

The x component of the force applied to the box is the same as it was before, 245 Newtons. The answer in this case is that the box starts to move. The horizontal force on the box is greater than the maximum value that the force of static friction can reach. Here, the y component of the force on the box is acting to reduce the normal force (and therefore the force of friction), rather than adding to mg and producing a higher normal force as seen in the earlier situation.

4-9 A block with a mass of 500 grams is pulled up an incline plane at a constant velocity by a force of 3.86 N. The plane has an angle of 53° with respect to the horizontal. What is the coefficient of friction between the block and the plane?

Solution: First, draw a diagram. In the vector diagrams that you draw for situations where there are many different forces on an object, it is a good idea to label the direction of motion. That way you can call any force that acts in the direction of motion positive and any force that acts against the direction of motion negative. We know that friction always acts to oppose the motion, so the force of friction is drawn and labeled f.

Insight

The tricky part about this diagram is getting straight which force is $mg\sin 33^o$ and which is $mg\cos 33^o$. The diagram in the box below is intended to clarify the situation. Perhaps

you can see the geometry of why the 33° is where it is in the diagram. If you're not interested in looking for similar triangles and worrying about the geometry on a test, you can just remember that *any time an object is on an incline plane the component of the*

Speed *gravitational force that acts parallel to the plane is sin.*

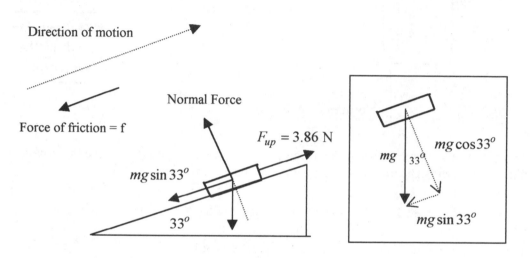

Another piece of important information given in this problem is that the block is pulled up the plane at a constant velocity. This means that all the forces acting up the plane must equal all the forces acting down the plane. Remember, constant velocity means that there is no acceleration. No acceleration means that there is no net force on an object. So therefore, we can write:

$$F_{up} = mg\sin 33^o + f \;\;\Rightarrow\;\; 3.86 \text{ N} = (0.5 \text{ kg})(9.8 \text{ m/s}^2)(0.545) + f$$

$$f = 3.86 \text{ N} - 2.67 \text{ N} = 1.19 \text{ N}$$

Do you remember the original question? It's OK if you don't. It is not unusual to have to go back and read the whole question again carefully at a point like this to make sure you are on the right track. The question was to find the coefficient of friction. Since we have the force of friction now, it looks like we're on the right track. The general form for the force of friction is $f = \mu \times$ normal force. Now we need to find the normal force. We are looking for all forces that act perpendicular to the plane or any force that has a component acting perpendicular to the plane. F_{up} acts parallel to the plane, so it does not contribute to the normal force. The normal force in this problem is the component of the gravitational force on the block that acts perpendicular to the plane. Look back at our diagram. The normal force is $F_{normal} = mg\cos33^o$. Now we can solve for the coefficient of kinetic friction:

$$f = \mu \times F_{normal} = \mu mg\cos33^o \quad \Rightarrow \quad 1.19N = \mu(0.5\text{ kg})(9.8\text{ m/s}^2)(.839)$$

$$\mu = \frac{1.19\text{ N}}{4.11\text{ N}} = 0.29\text{ N}$$

Now look carefully at just this last equation. Is the answer correct? The answer is not 0.29 N, it is just **0.29**! The unit Newton is both in the numerator and denominator of that fraction. Newton divided by Newton is a dimensionless quantity.

Watch
Out!

Let's think about the physics of μ for a moment. Having a coefficient of friction that is 0.29 says that 29% of an object's weight is the force of friction that an object will experience sliding over this particular surface. If the object weighs 100 N, then the force of friction will be 29 N. If the object weighs 10 N then the force of friction will be 2.9 N. So now that you see what μ really stands for, do you also see that it is not an absolute number of Newtons? It is not a force. It is a fraction. Remember that mathematically speaking if you divide Newtons by Newtons, pounds by pounds, seconds by seconds…etc…you will get a dimensionless quantity.

5

TORQUE AND EQUILIBRIUM

Things to Remember

- The definition of **torque** is $\tau = F \times l$ where l is called the lever arm. See problem 5-1 for a description of the lever arm.

- The formula for **center of gravity** in one dimension is $x_{cg} = \dfrac{\sum m_i x_i}{\sum m_i}$. See problems 5-2 and 5-3.

- The first condition for equilibrium is that there is no net force on an object. The object may be either at rest or moving at a constant velocity. Remember that net forces produce accelerations and that if an object is moving at a constant velocity there is no net force on it. The mathematical formulas for the above statements in the x and y dimensions are $\sum F_x = 0$ and $\sum F_y = 0$.

- The second condition for equilibrium is that the sum of the torques is equal to 0, or $\sum \tau = 0$.

Torque

5-1 A wrench is being used to loosen a nut as shown. If a force of 20 Newtons is applied to the wrench, what is the torque on the nut?

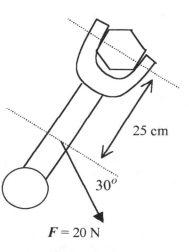

Solution: The general formula for torque is $\tau = force \times l$. We can't just use 20 N for the force, however. **The force and the lever arm must be at a 90° angle** in order for this formula to work. Taking the component of the 20 N force which is at 90° to the lever arm,

Remember

34

$$\tau = 20\,\mathrm{N}\cos(30^{o})(.25\mathrm{m}) = 4.33\,\mathrm{N \cdot m}$$

Center of Gravity

The "center of gravity" can also be thought of as center of mass. For a symmetrical object like a perfectly round cannon ball, the center of gravity would be at the exact center of the sphere. If a sphere were constructed of half aluminum (light) and half lead

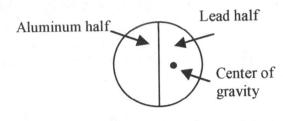

as shown, then the center of gravity would be shifted toward the heavier part of the sphere. Some problems ask you to find the center of gravity for several independent masses. For these problems you need to know the mass and position of each object.

5-2 Find the center of gravity for the arrangement of masses shown.

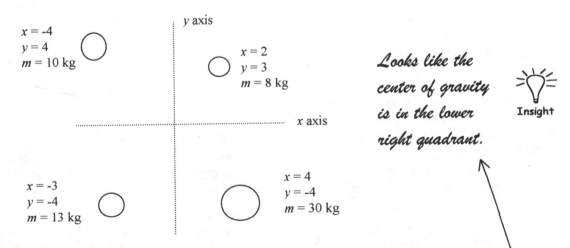

Solution: You may be trying to memorize formulas and learn some concepts for the first time, but don't let your mind be so preoccupied that you never just sit back, look at the problem and use your intuition. Look at the masses and distances here. Does it look like the center of gravity should end up being below the *x* axis? Look a little farther. That 30 kg mass in the bottom right quadrant looks like it is going to dominate the calculation doesn't it? If you see this at the beginning of a problem, you may want to write it down next to the problem just as a reminder.

$$x_{cg} = \frac{\sum m_i x_i}{\sum m_i} = \frac{10 \text{ kg}(-4) + 8 \text{ kg}(2) + 13 \text{ kg}(-3) + 30 \text{ kg}(4)}{10 \text{ kg} + 8 \text{ kg} + 13 \text{ kg} + 30 \text{ kg}} = 1.15$$

$$y_{cg} = \frac{\sum m_i y_i}{\sum m_i} = \frac{10 \text{ kg}(4) + 8 \text{ kg}(3) + 13 \text{ kg}(-4) + 30 \text{ kg}(-4)}{10 \text{ kg} + 8 \text{ kg} + 13 \text{ kg} + 30 \text{ kg}} = -1.77$$

Remember to keep the x's and y's separate when calculating the center of gravity. The final answer is what we expected. The center of mass is in the lower right quadrant at the point $x = 1.15$, $y = -1.77$.

5-3 Where is the center of gravity of this solid object? It is made of a material that has uniform density.

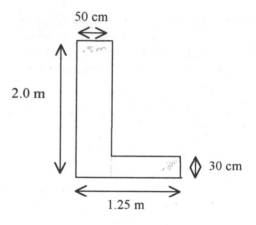

50 cm

.5 m

2.0 m

30 cm

1.25 m

Solution: First we need to pick a point as a reference. Let's pick the bottom left corner of this L shape. The strategy here is first to reduce the solid shapes to two point masses. Then we can calculate center of gravity the same way we did in the previous problem.

Insight

The "fly in the ointment" for this problem is that we don't know the density of the object or anything about it's mass. We just know that it has uniform density. Learn to suspect that when it seems as if you need a quantity to solve a problem, it is probably going to cancel out somewhere in the calculation. So keep in mind that we are looking for an opportunity to create an equation where mass or density cancels out. Density (ρ) is equal to mass/volume. In this case we can use mass/area since we only have two dimensions.

The center of mass for the vertical rectangle is $x = 0.25$ m, $y = 1.0$ m. We know that the center of mass is in the middle of both the x any y dimensions of the vertical rectangle.

Watch Out!

The horizontal rectangle at the bottom now **does not include the corner portion shared by both rectangles.** The surest way to get burned on this problem is to count that corner piece shared by both rectangles twice.

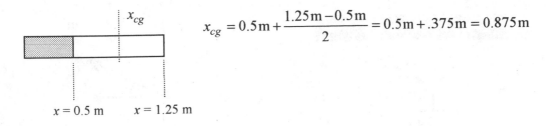

$$x_{cg} = 0.5\text{m} + \frac{1.25\text{m} - 0.5\text{m}}{2} = 0.5\text{m} + .375\text{m} = 0.875\text{m}$$

$x = 0.5$ m $x = 1.25$ m

The value for y_{cg} is just 0.15 m.

We now know the positions of the two point masses that represent the rectangles. The formula for center of gravity involves mass, but we can just as easily use area to "weight" the distances. The area of the vertical rectangle is $2.0\text{ m} \times 0.5\text{ m} = 1.0\text{ m}$. The area of the horizontal rectangle is $0.75\text{ m} \times 0.30\text{ m} = 0.225\text{ m}$. We can now redraw the problem using point masses and areas as shown.

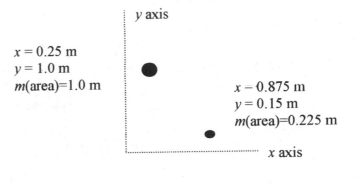

$$x_{cg} = \sum \frac{area_i x_i}{area_i} = \frac{0.25\text{ m}(1.0\text{ m}) + 0.875\text{ m}(0.225\text{ m})}{1.225\text{ m}} = 1.18\text{ m}$$

$$y_{cg} = \frac{1.0\text{ m}(1.0\text{ m}) + 0.15\text{ m}(0.225\text{ m})}{1.225\text{ m}} = 1.03\text{ m}$$

Equilibrium Problems

5-4 Hang a 50 kg mass with ropes making angles of 30° and 45° as shown below. Calculate the tension in the ropes.

Solution: **Note that all the forces come together at the knot in the rope.** Draw a force diagram about this point. The only laws to apply are for equilibrium in the x and y directions.

Insight

$\sum F_x = 0:$ $T_1 \cos 30^o = T_2 \cos 45^o$ $\sum F_y = 0:$ $T_1 \sin 30^o + T_2 \sin 45^o = 294\,\text{N}$

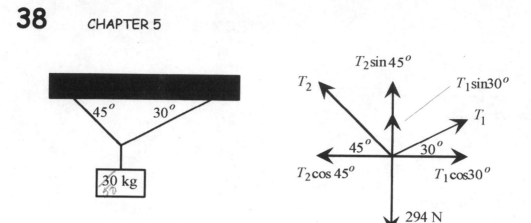

This provides two equations in two unknowns. Because $\sin 45^o = \cos 45^o$ rewrite

$$T_1(\sin 30^o + \cos 30^o) = 294\,\text{N} \qquad \text{or} \quad T_1 = 215\,\text{N}$$

and

$$T_2 \cos 45^o = (215\,\text{N})\cos 30^o \qquad \text{or} \qquad T_2 = 263\,\text{N}$$

5-5 In the arrangement shown below, what is the minimum coefficient of friction to prevent the 8.0 kg mass from sliding?

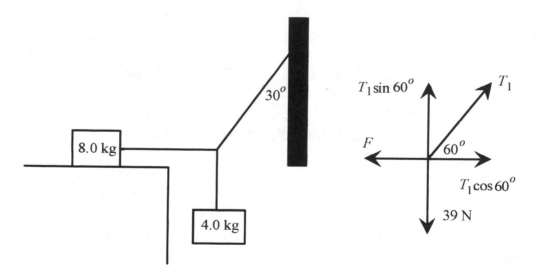

Solution: The forces come together where the ropes come together. This point is not moving so apply the equilibrium conditions here. Again, draw a separate vector diagram **Pattern** (at right above) just as in the previous problem.

The equilibrium conditions are:

$$\sum F_x = 0: \quad T_1 \cos 60^o = F \qquad \sum F_y = 0: \quad T_1 \sin 60^o = 39\,\text{N}$$

Solving for T_1: $\quad T_1 = \dfrac{39\,\text{N}}{\sin 60^o} = 45\,\text{N}$

The force is: $\quad F = 45\,\text{N}\cos 60^o = 22.5\,\text{N}$

This force is supplied by the frictional force μmg so $\quad \mu = \dfrac{22.5\,\text{N}}{8.0\,\text{kg}\cdot 9.8\,\text{m/s}^2} = 0.29$

This is the minimum coefficient of friction that will prevent the system from moving.

5-6 A 500 N diver is on the end of a 4.0 m diving board of negligible mass. The board is on pedestals as shown below. What are the forces that each pedestal exerts on the diving board?

Solution: There are no forces in the x direction but in the y direction the pedestal at A and the diver are acting down while the force at pedestal B is acting up. Pedestal B is in compression while pedestal A is in tension. Draw the vector diagram and write the equations.

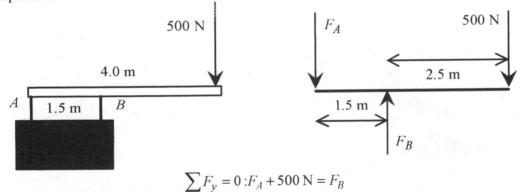

$$\sum F_y = 0 : F_A + 500\,\text{N} = F_B$$

In this problem apply $\sum \tau = 0$ about pedestal A. Taking the rotation point at pedestal A, the diver produces a clockwise torque, and pedestal B an anti-clockwise torque that are in equilibrium. Note that this is not the only possible rotation point.

$$\sum \tau = 0\,500\,\text{N}\cdot 4.0\,\text{m} = F_B \cdot 1.5\,\text{m}$$

Solve for F_B: $\quad F_B = \dfrac{500\,\text{N}\cdot 4.0\,\text{m}}{1.5\,\text{m}} = 1330\,\text{N}$

and solve for F_A: $F_A = F_B - 500\,\text{N} = 1330\,\text{N} - 500\,\text{N} = 830\,\text{N}$

As an exercise write the $\sum \tau = 0$ about pedestal B and the diver.

5-7 Place a 7.0 m uniform 150 N ladder against a frictionless wall at an angle of 75°. What are the reaction forces at the ground and wall and the minimum coefficient of friction of the ground?

Solution: The diagram below shows the ladder with the 150 N acting down at the center of the ladder and the sides of the triangle formed by the ladder, wall, and ground. The vector portion of the figure shows the two components of the **reaction force at the ground. In ladder problems remember about the reaction force at the ground.** Also, the reaction force at the ground is not necessarily in the direction of the ladder. This (vector) force can be written in terms of a force (magnitude) at an angle (with horizontal and vertical components) or directly in component form as done here. Writing the force in component form helps to avoid the temptation to place the force at the same angle as the ladder. Either way requires two variables. Since the wall is frictionless, the reaction at the wall is entirely horizontal.

Insight

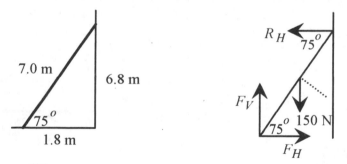

The equilibrium conditions are:

$$\sum F_x = 0: \quad F_H = R_H \qquad \sum F_y = 0: \quad F_V = 150\,\text{N}$$

$\sum \tau = 0$: The torque on the ladder is taken about the point where the ladder contacts the ground. Note that this choice eliminates two variables from the torque statement. Torque is the (component of the) force at **right angles to the lever arm** times that lever arm. The 150 N weight has a component $150\,\text{Ncos}75^o$ times 3.5 m. The reaction at the wall has a component $R_H \cos 15^o$ times 7.0 m so

$$(150\,\text{Ncos}75^o)3.5\,\text{m} = (R_H \cos 15^o)7.0\,\text{m} \quad \text{or} \quad 150(0.90)\,\text{N} \cdot \text{m} = R_H(6.8)\,\text{N} \cdot \text{m}$$

$$R_H = (75\text{N})(\cos75^o/\cos15^o) = 20\text{N} \quad \text{and} \quad F_H = 20\text{N}$$

F_V and F_H are related via the coefficient of friction, $F_H = \mu F_V$ so $\mu = 20/150 = 0.13$.

Second Solution: There is another way to calculate the torques.   Notice that the 150 N force is horizontally 0.90 m away from the pivot point. If the 150 N force vector is moved vertically so its tail is on a horizontal line from the pivot point then the torque is easily written as $(150\,\text{N})(0.90\,\text{m})$. A similar operation can be performed on R_H yielding a torque $R_H (6.8\,\text{m})$. Vectors maintain their length (magnitude) and orientation (angle) but **can be moved about a vector diagram for our convenience.** If the torque statement is written in this way it is identical to the torque statement in the first solution:

$$150\,\text{N} \cdot 0.90\,\text{m} = R_H 6.8\,\text{m}$$

6

CIRCULAR MOTION

Things to Remember

- The period (T) of circular motion is the amount of time that it takes to complete one revolution. The velocity can be related to the period by the formula $v = 2\pi r / T$.
- The radius for an object in circular motion above the earth is the distance from the center of the earth. This means that if you are given the distance above the surface of the earth, you need to add the radius of the earth to that number.
- When doing circular motion problems, keep in mind that you can work the problem from two different reference frames: a reference frame at rest and the reference frame of the object being rotated.
- Centripetal acceleration is v^2 / r.

6-1 The space shuttle is orbiting the earth at a height of 350 km above the surface of the earth. What must the speed of the space shuttle be in order to maintain a steady orbit?

There are two ways to approach this problem. For now, let's not worry about the numerical value of the solution. First, let's just consider two ways of thinking about this situation.

Solution A: From a reference frame at rest relative to the circular motion. From the perspective of a fixed reference point outside the motion, we see the space shuttle going in circles around the earth. The (direction of) velocity of the shuttle is changing. Remember that the speed is constant but if the direction changes then the velocity changes because velocity is a vector, while speed is not. Now if the velocity changes there must be an acceleration. Just think about linear motion. If a car increases its velocity it must accelerate, right? The same goes for a constant speed, but changing direction. The velocity is changing (at a constant rate) and so there is a constant acceleration. The direction of that acceleration is toward the center of the circle (center of the earth).

OK, since we know that there has to be a net acceleration, then we also know there must be a net force on the shuttle directed toward the center of the earth. At this point you may think, "Of course, the force of gravity acts toward the center of the earth." Yes that's correct, but that alone won't solve the problem for us. What we can write so far is the following (remembering that this must be a big G problem, not a little g problem because we are not close to the surface of the earth):

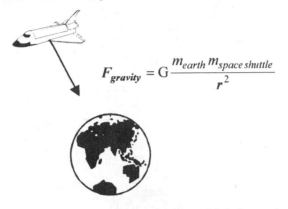

$$F_{gravity} = G \frac{m_{earth}\, m_{space\, shuttle}}{r^2}$$

The question asked is of the orbital speed of the shuttle, which is not in the formula for the force of gravity. The other principle that we need here is that of centripetal force. The way this will be presented from the reference frame outside the motion is that the centripetal force is center-directed and always equal to mv^2/r. In this case the m is the mass of the shuttle. The m in mv^2/r always refers to the mass of the object that is moving in the circle. **Operationally what this means is this: When doing circular motion problems, find the net center-directed force (sometimes you may have to add or subtract forces to get the net force). Then set this net center-directed force equal to** mv^2/r.

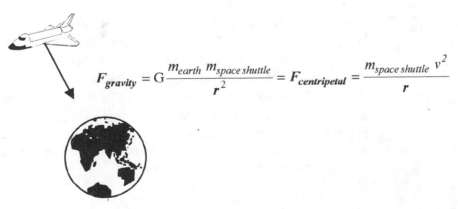

$$F_{gravity} = G \frac{m_{earth}\, m_{space\, shuttle}}{r^2} = F_{centripetal} = \frac{m_{space\, shuttle}\, v^2}{r}$$

Solution B: From the reference frame of the astronaut inside the space shuttle. Have you ever seen a video of space shuttle astronauts? They are weightless. They float freely about the shuttle without any effects on them from the force of gravity. There are many scenes on TV and IMAX movies where they pass food or other objects back and forth in a weightless environment. Now, the force of gravity is acting on those astronauts just as it is acting on the shuttle. However from their frame of reference, moving in a circle, there is a force (referred to as centrifugal force) that counterbalances the force of gravity. From their point of view, the diagram of the problem would look like this:

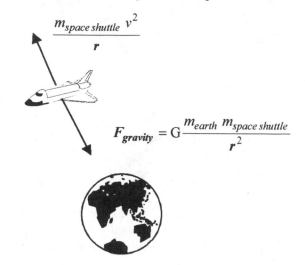

Since the reference frame is the circular motion, there is no net force on the shuttle. There is no movement in the radial direction away from the earth or closer to it. Therefore all the forces acting radially outward are equal to all the forces acting radially inward. And again

we have $F_{gravity} = G\dfrac{m_{earth}\, m_{space\,shuttle}}{r^2} = \dfrac{m_{space\,shuttle}\, v^2}{r}$.

Contrary to the implications of many textbooks and physics instructors, there is **nothing wrong with thinking about the problem from this point of view**. A word of caution however, if your professor has a real hang-up with thinking about the problem this way and looks like the character pictured here when he or she sees the above diagram showing mv^2/r acting radially out, then you may want to stick to doing things as we've shown in solution A. If your professor is more open minded, then choose whichever method you feel more comfortable with.

Proceeding now with the math on this problem,

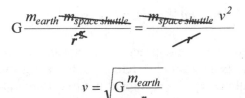

$$v = \sqrt{G\frac{m_{earth}}{r}}$$

Now it is very important at this point to remember that r is the **distance from the center of the earth.** We were given that the space shuttle was 350 km above the surface of the earth. We need to add that to the radius of the earth which is 6.38 million meters.

Watch Out!

$$r = 6.38 \times 10^6\,\text{m} + 350 \times 10^3\,\text{m} = 6.38 \times 10^6\,\text{m} + 0.35 \times 10^6\,\text{m} = 6.73 \times 10^6\,\text{m}$$

Now, solving for the velocity,

$$v = \sqrt{6.67 \times 10^{-11}\,\text{N} \cdot \text{m}^2/\text{kg}^2 \frac{5.98 \times 10^{24}\,\text{kg}}{6.73 \times 10^6\,\text{m}}} = 7698\,\text{m/s} \Rightarrow 7700\,\text{m/s}$$

You may not have time on a test to convince yourself completely that the units really do work out to m/s, but by looking at the units of the gravitational constant you can see that you need to use meters for the radius and kg for the mass (not km for radius, for example). In this case, a quick glance at the units of the constant in the equation tips you **Speed** off as to what the units of the variables are.

6-2 A child playing on a swing has a velocity of 3 m/s at the very bottom of the path of the swing (when his velocity is exactly parallel to the ground). How much heavier does the child feel if her mass is 50 kg and the length of the swing (and the radius of the circular motion) is 3 meters? Express the total weight in Newtons.

Solution: The force of gravity acts on the child and is always the same. It is $F_{gravity} = 50\,\text{kg} \times 9.8\,\text{m/s}^2 = 490\,\text{N}$. Since the question is asking about what the child feels, we use the thinking in the manner of solution B above. To her there is a force mv^2/r that acts radially out and adds to her weight. That force, and the total apparent weight of the child may be calculated as follows:

$$mv^2/r = 50\,\text{kg} \times (3\,\text{m/s})^2/3\,\text{m} = 150\,\text{N}$$

$$\text{Total Apparent Weight} = 450\,\text{N} + 150\,\text{N} = 600\,\text{N}$$

Futurists have considered using this effect in the design of space stations. The movie "2001: A Space Odyssey" is an example. Here a giant rotating space station is shown with the force of gravity simulated by the rotational motion. Remember the elevator problem from the last chapter where Einstein appeared and explained that an acceleration can produce the same effect as gravity? That was for the case of vertical acceleration in an elevator, but the same effect can be produced with circular acceleration.

6-3 A man is whirling a ball on the end of a cord around his head in a horizontal circle (parallel to the ground). The period of rotation is 1.0 s and the radius 1.0 meters. What is the tension in the cord?

Solution: The force that makes the ball travel in a circle must be equal to mv^2/r. The velocity of the mass is the distance travelled (one circumference) divided by the time to travel one circumference (that time is called the period, T):

Remember

$$v = \frac{2\pi r}{T} = \frac{2\pi \cdot 1.0\ \text{m}}{1.0\,\text{s}} = 6.28\,\text{m/s}$$

The force or tension in the cord is

$$F = \frac{mv^2}{r} = \frac{2.0\,\text{kg}\,(6.28\ \text{m/s}^2)}{1.0\ \text{m}} = 79.0\ \text{N}$$

Now let's think about how the tension in the cord is interpreted. If you are whirling the cord, the tension, from your point of view, is radially out. From the point of view of the mass in the sling, however, the tension is pointing in. Take a "thought trip" in an airplane flying in a horizontal circle with its wings vertical. A force meter between you and the seat would read a number equal to your mass times v^2/r. The force acting on you is toward the center of the circle but you exert an equal and opposite force on the seat that is radially out of the circle (this is analogous to the normal force concept). In some airplanes, if the banking is done too hard this radially outward force can be so great as to cause the blood in the pilot to remain in the lower part of the body resulting in a loss of oxygen (carried by the blood) to the brain and "blackout".

6-4 Suppose that now the ball is whirled in a vertical circle at a constant velocity of 6.28 m/s. Calculate: (a) the tension in the cord at the top of the motion, (b) the tension in the cord at the bottom of the motion, and (c) the minimum velocity to keep the mass from "falling out" at the top of the circle.

Solution: First let's diagram this situation and label the forces at the different stages of the motion.

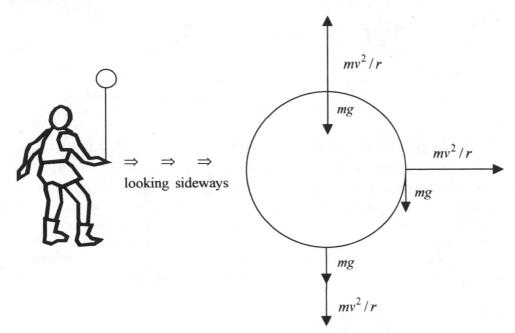

mv^2/r

mg

mv^2/r

mg

looking sideways

mg

mg

mv^2/r

We know that the mv^2/r force is 79 N from the previous problem. Now let's find the gravitational force mg.

$$mg = 2.0\,\text{kg}\,(9.8\ \text{m/s}^2) = 19.6\ \text{N}$$

(a) The tension in the cord at the top of the motion is 79.0 N − 19.6 N = 59.4 N
(b) The tension in the cord at the bottom of the motion is 79.0 N + 19.6 N = 98.6 N

(c) The minimum velocity required can be found by considering the condition where $mv^2/r \le mg$. In this case, at the top of the motion, the mass will fall out of the loop. So let's set the two forces equal and solve for the minimum velocity needed to keep the mass in the loop.

$$\frac{mv_{min}^2}{r} = mg \implies v_{min} = \sqrt{rg} = \sqrt{1.0\ \text{m}\,(9.8\ \text{m/s}^2)} = 3.13\ \text{m/s}$$

7

WORK AND ENERGY

Things to Remember

- Kinetic energy is the energy an object has due to its motion. The faster and more massive an object, the greater the kinetic energy. The formula is: $KE = (1/2)mv^2$.
- Gravitational potential energy is the energy that an object has due to its position or height. The greater the mass and the greater the height off the ground, the greater the gravitational potential energy. A 5 gram marble dropped in the sand on the beach from a height of 1 meter will make less of a crater in the sand than a 10 pound shot put dropped from an airplane 1000 meters high. The formula for gravitational potential energy is $GPE = mgh$. Some books label gravitational potential energy simply as PE instead of GPE.
- The basic formula for work is $Work = Force \times Displacement$. This often appears as $W = Fs$ (s is used for displacement). Keep in mind that you must use the component of the force that acts in the direction of the displacement.
- The units of energy (work has the same units) is Joules or $Newton \cdot meters$. Power is energy/time or J/s = Watts.

7-1 (a) A woman weighing 700 N rides an elevator up two levels. The elevator rises 15 meters. (b) On the trip back down the woman is joined by two friends. They all have a total weight of 2000 N. The elevator descends 15 meters. Assuming that the elevator moves at constant velocity in both cases, how much work is done by the elevator on the passengers in each case?

Solution: (a) Assuming that the elevator moves up at a constant velocity for 15 meters means that there was no net force on the 700 N woman for those 15 meters. In other words the force of the elevator on her balances the force of gravity. So we have a 700 N force acting in the same direction as a displacement of 15 meters. The work done by the elevator is $700 \text{ N} \times 15 \text{ m} = 10,500 \text{ J}$.

(b) For the second case where the elevator is descending, the force again is equal to the weight of the passengers, this time 2000 N, but the displacement is in the opposite direction of the force. For this situation we need to include a **minus sign** to indicate this: $2000 \text{ N} \times (-15 \text{ m}) = -30,000 \text{ J}$. **Remember to use a minus sign for the value of work when the force acts one way and the displacement is in the opposite direction.** Remember

7-2 A block slides down a frictionless 58 degree incline plane with a slant length of 4 m. The block has a mass of 3 kg. What is the velocity at the bottom of the plane?

Solution: A way to understand this problem is that at the top of the plane, the block has gravitational potential energy (mgh). This energy "goes into" kinetic energy, or the energy of motion ($(1/2)mv^2$). Remember this "goes into" concept. It will help you keep minus signs straight in thinking about these problems. "Goes into" may be shortened to the one descriptive word **goesinto**. As the problems get more complicated you will find this concept more and more useful. Insight

Now let's diagram the problem and find the height of the block off the ground.

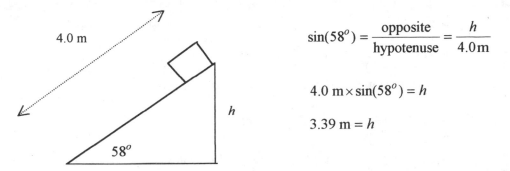

$$\sin(58^{\circ}) = \frac{\text{opposite}}{\text{hypotenuse}} = \frac{h}{4.0\text{m}}$$

$$4.0 \text{ m} \times \sin(58^{\circ}) = h$$

$$3.39 \text{ m} = h$$

The gravitational potential energy of the block at the top of the plane is calculated as follows:

$$mgh = (3.0 \text{ kg})(9.8 \text{ m/s}^2)(3.39 \text{ m}) = 99.66 \text{ Joules}$$

Now we know how much energy we start with. The next question is what does this energy go into? Since the block has no friction, all the gravitational potential energy **goesinto** kinetic energy.

$$99.7 \text{ J} = (1/2)mv^2 \implies v = \sqrt{99.7 \text{ J} \times 2/3.0 \text{ kg}} = 8.15 \text{ m/s}$$

7-3 Let's add a twist to the previous problem. Suppose that we have the same situation, except that the plane is not frictionless. Let's say that the coefficient of kinetic friction is $\mu = 0.25$. Now calculate the velocity at the bottom of the plane.

Solution: In this case we start out with the same amount of energy as before, 99.66 J. However in this case the energy **goesinto** two places. It goes into kinetic energy as before and it goes into work to overcome the force of friction that opposes the motion of the block down the plane. Using the **goesinto** as an equal sign, we can write:

$$\text{potential energy} = \text{kinetic energy} + \text{work vs. friction}$$

$$99.7\,\text{J} = (1/2)mv^2 + Force_{friction} \times 4.0\,\text{meters}$$

Remember to keep things straight here. 4.0 meters is the displacement along which the force of friction acts. It belongs in the formula for work. The value of 3.39 meters is what we calculated earlier to be the height off the ground. That belongs in the formula for potential energy.

In order to calculate the force of friction, remember the definition of friction force is $\mu \times normal\,force$. The normal force is $mg\sin(58°)$ (see problem 4-9). So,

$$Force_{friction} = 0.25(3.0\,\text{kg})(9.8\,\text{m/s}^2)(\sin 58°) = 6.23\,\text{N}$$

The work done to overcome friction is then $6.23\,\text{N} \times 4.0\,\text{m} = 24.9\,\text{J}$.

Now we can plug this into the original energy equation and solve for velocity.

$$99.7\,\text{J} = (1/2)mv^2 + 24.9\,\text{J} \implies v = \sqrt{74.8\,\text{J}(2)/(3.0\,\text{kg})} = 49.9\,\text{m/s}$$

Now, pause and think about this answer. Look back at the answer to the previous problem. Does the number here look about right? Think about this before reading any further.

**Watch
Out!**

Will the velocity in this problem be greater than or less than the velocity in the previous case? Since the energy we started with has to go into overcoming friction as well as into kinetic energy, we should end up with less kinetic energy than we did in the first problem and less velocity. But the number here is greater, much greater. We must have made a mistake. Actually we just forgot to take the square root, so **the correct answer is 7.06 m/s. Remember to think about the numbers you are getting as you are going through the problem and make sure that they make sense.**

**Watch
Out!**

Beware of the following formula that is highlighted in many textbooks and called the "Work-Energy Theorem".

$$W = \Delta KE = KE_{final} - KE_{initial} = (1/2)mv^2_{final} - (1/2)mv^2_{initial}$$

If you see this formula in your text, please realize that **this is not a general formula that works in all types of problems**. You will notice that there is no potential energy in this formula. Later in your text it will probably be emphasized that a more general formula is:

$$W = \Delta KE + \Delta PE$$

Even this can be misleading unless you know all the "rules" of how to play games with minus signs in these types of problems.

For example, suppose that we are given a problem that says, "A girl puts 10 J of work into pushing a ball up a frictionless 30° incline plane. What is the kinetic energy at the top?" In this case we start with the work and that energy **goesinto** a change in kinetic energy plus a change in potential energy. In that case we could use the above formula with no strings attached. However suppose we changed the problem so that the girl pushed the ball down the plane doing 10 J of work. The velocity of a ball pushed down an incline plane will be faster than the velocity of the ball pushed up the plane. In this case both the work and the potential energy **gointo** kinetic energy and we would use the equation:

$$W + \Delta PE = \Delta KE$$

Suppose we are given a problem where a ball is dropped off a cliff from a height of 100 meters and we are asked how much kinetic energy it has when it hits the ground. In this case we start with potential energy and that energy **goesinto** kinetic energy when the ball hits the ground. So for that situation we have:

$$\Delta PE = \Delta KE$$

By the way, don't think that all possible formulas for types of energy will be covered in the same chapter in your text. For example, many texts don't talk about springs until after the chapter on work and energy. There is a unique formula for potential energy in a coiled spring. It is dangerous to simply memorize one "work-energy theorem" equation and try to use it in all circumstances. If you choose this route, be sure that you have also memorized when each quantity in the equation is positive and when it is negative. We believe that you are better off thinking about the problem in terms of energy flow. How much energy do you start with and where does it go?

Watch
Out!

7-4 A car is speeding along and the driver suddenly applies the brakes as hard as he can. The car leaves skid marks for a length of 50 meters. If the coefficient of friction between the tires and the road is 0.4, how fast was the car going at the moment the brakes were applied?

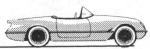

Solution: OK, let's think through this.

- The question asks how fast the car was going. Velocity is in the kinetic energy equation, $(1/2)mv^2$.
- We know the length of the skid marks.
- If we knew the force that opposed the car, the force of friction, we could use $Work = Force \times Distance$. This would give us the work done to overcome friction.

So our thinking so far is that the kinetic energy that the car has before the brakes are applied **goesinto** work against friction.

$$(1/2)mv^2 = F_{friction} \times 50 \text{ meters}$$

We need $F_{friction}$ now. Look back at the question and answer to yourself how we are going to get $F_{friction}$.

Remember

Do you remember that $F_{friction} = \mu \times \text{normal force}$? Remember this. It is one of those things that will keep coming up for a long time. The normal force is just mg, the weight of the car. OK, now we can look back at the question and get the mass of the car. Look back now for the mass of the car.

The mass of the car is not given in the question! When confronted with a situation like this on an exam, many things can go through a student's mind.

Let's plug μmg in for $F_{friction}$ in the equation above and see what happens.

$$(1/2)mv^2 = \mu mg \times 50 \text{ meters}$$

Do you see it now? The mass, m, is on both sides of the equation, so it cancels out. Please remember this. If it looks like you really need to be given a certain value in order to answer a question, then maybe it cancels out somewhere.

Insight

$$v = \sqrt{2 \times 0.4 \times 9.8 \text{ m/s}^2 \times 50 \text{ m}} = \sqrt{588 \text{ m}^2/\text{s}^2} = 19.8 \text{ m/s}$$

Notice also that $\sqrt{\text{m}^2/\text{s}^2} = \text{m/s}$. We take the square root of the units as well as the number 588.

7-5 A solar panel that converts sunlight into electricity is capable of producing an average of 1000 Watts of power for 10 hours per day. The panel is guaranteed for 10 years. What must be the cost of the solar panel if it is to be as cheap as an electric bill of 10 cents per Kilowatt·hr ?

Solution: The best place to start this problem is with the units of power and energy. Remember that Power = Energy/time, and Watts = Joules/seconds. Can you see that Kilowatt·hr must be a unit of energy? If Watts = Joules/seconds, then Joules = Watts × seconds. So Kilowatt·hr looks like Watts × seconds in so far as they are both power × time. Kilowatt·hr is not equal to Joules, but it still can be converted to Joules, and is a unit of energy.

This means that an electric bill is based on how much energy we use, not how much power. This makes sense, doesn't it? You could turn on a 100 Watt light bulb for 10 seconds or you could turn it on for 10 days. You're going to pay for power × time, or energy, not power.

OK, now to the question. Let's find out how much total energy that solar panel can generate in 10 years if it can produce on average 1000 W for 10 hours per day. This last part, 1000 W for 10 hours per day, is 10 Kilowatt·hr for each day. Ten years multiplied by that is

$$\frac{10 \text{ KW} \cdot \text{hr}}{\cancel{\text{day}}} \times \frac{10 \cancel{\text{years}}}{1} \times \frac{365 \cancel{\text{days}}}{1 \cancel{\text{year}}} = 36,500 \text{ KW} \cdot \text{hr}$$

At 10 cents per Kilowatt·hr , this would cost us the following amount:

$$36,500 \cancel{\text{KW} \cdot \text{hr}} \times \frac{10 \cancel{\text{cents}}}{\cancel{\text{KW} \cdot \text{hr}}} \times \frac{1 \text{ dollar}}{100 \cancel{\text{cents}}} = 3650 \text{ dollars}$$

So in order to be cheaper than our electric bill, this solar panel would have to cost less than $3650. If you are in the business of making solar panels, you have to be able to manufacture them for even less in order to make a profit.

7-6 A car accelerates uniformly from 0 to 60 mi/hr in 6.0 seconds. If the car weighs 2000 lbs, how much power is needed during the acceleration in horsepower (hp)? Ignore friction.

Solution: We know that power = energy/time. We have the time, 6.0 seconds. Think over the different forms of energy (work, kinetic, potential). The total energy that the car has after the 6 seconds is kinetic energy $(1/2 mv_f^2)$. The final velocity of the car is:

$$v_f = \frac{60 \text{ mi}}{\text{hr}} \times \frac{1.609 \text{ km}}{1 \text{ mi}} \times \frac{1000 \text{ m}}{\text{km}} \times \frac{1 \text{ hr}}{60 \text{ min}} \times \frac{1 \text{ min}}{60 \text{ sec}} = 26.8 \text{ m/s}$$

$$\text{and} \quad 2000 \text{ lbs} \times \frac{1 \text{ kg}}{2.205 \text{ lbs}} = 907 \text{ kg}$$

$$KE = (1/2)(907 \text{ kg})(26.8 \text{ m/s })^2 = 3.26 \times 10^5 \text{ J}$$

$$P = 3.26 \times 10^5 \text{ J} / (6 \text{ sec}) = 5.43 \times 10^4 \text{ Watts} \times \frac{1 \text{ hp}}{745.7 \text{ W}} = 72.8 \text{ hp}$$

In reality it would take an engine with close to 300 hp to accelerate a car from 0 to 60 in 6 seconds. Remember that we ignored friction. Much of the power from an engine goes into mechanical friction, fighting air resistance, and friction between the tires and the road. What we have calculated above is the amount of power that goes into producing velocity.

8

IMPULSE AND MOMENTUM

Things to Remember

- Impulse = average force$\times \Delta t$, where Δt is the amount of time over which the force acts. In most problems we assume that a constant force acts for a certain amount of time. The SI units are $N \cdot s$.

- $p = mv$ is the definition of momentum. Remember that momentum, as velocity, is a vector. The SI units of momentum are $kg \cdot m/s$.

- The "impulse-momentum theorem" states that impulse equals change in momentum or $F \Delta t = \mu mv = mv_{final} - mv_{initial}$.

- Momentum is always conserved in a collision and kinetic energy is sometimes conserved. An **elastic** collision (billiard balls) is one where kinetic energy is conserved. An **inelastic** collision (lumps of clay) is one in which kinetic energy is not conserved. The objects stick together in an inelastic collision.

8-1 A golfer hits a golf ball weighing 45 grams off a tee. A high-speed camera is able to measure that the duration of impact of the club with the ball was 10 milliseconds. The ball is given a velocity after impact of 30 m/s. What was the average force applied by the club during the impact?

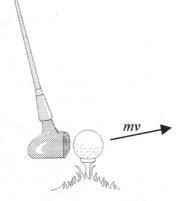

Solution: We use the "impulse-momentum theorem," $F \Delta t = \Delta mv$. The ball is initially at rest, so the initial momentum is 0. The final momentum is

$$45 \text{ grams} \times \frac{1 \text{ kg}}{1000 \text{ grams}} \times 30 \text{ m/s} = 1.35 \text{ kg} \cdot \text{m/s}$$

$$F(10 \times 10^{-3} \text{s}) = 1.35 \text{ kg} \cdot \text{m/s} \implies F = 135 \text{ N}$$

8-2 A basketball player bounces a basketball off the floor by giving it a downward velocity of 3.0 m/s at a height of 1 meter above the floor. The mass of the ball is 5 kg. The ball bounces off the floor and rises to a height of 0.80 meters (where it has 0 velocity). What is the magnitude and direction of the impulse applied to the ball by the floor?

Solution: We need to find the velocity of the ball just before and just after collision with the floor. The velocity of the ball is not 3.0 m/s just before the collision. It is 3.0 m/s at a height of 1 meter. Gravity then acts on the ball, accelerating it down. To find the velocity of the ball at the ground, we need to use one of those equations of motion that we saw earlier in chapter 2. You should have these memorized or written down as important equations to remember. These equations were:

Blast from the past

Equations	Things we know
2-1. $v = v_0 + at$	$v_0 = 3$ m/s
2-2. $x = v_0 t + (1/2)at^2$	$v = ?$
2-3. $x = (1/2)(v_0 + v)t$	$a = 9.8$ m/s^2
2-4. $v^2 = v_0^2 + 2ax$	$x = 1$ meter

We need to use the last equation, $v^2 = v_0^2 + 2ax$.

$$v^2 = (3 \text{ m/s})^2 + 2(9.8 \text{ m/s}^2)1 \text{ m} \implies v_{before} = 5.34 \text{ m/s}$$

We are calling this v_{before} because it is the velocity of the ball just before it hits the ground. Concerning minus signs in the above equation, keep in mind that the initial velocity is downward and the acceleration is downward. Therefore these values should have the same sign. If you like to call down negative then $v_{before} = -5.35$ m/s. We will stick with positive down and negative for the up direction. Now we need to find the velocity of the ball as it leaves the floor and is on its way back up. We know that it rises to a height of 0.8 meters and has 0 velocity when it reaches that height. Therefore, we know that $v = 0$ (not $v_0 = 0$). For this segment of the problem we are going to call v_0 the initial velocity of the ball on its way back up from the collision. So for now $v_0 = v_{after}$.

$x = 0.8$ meters $a = 9.8$ m/s^2

Again using the same equation,

$$0^2 = v_{after}^2 + 2(9.8 \text{ m/s}^2)(-0.8 \text{ m}) = v_{after}^2 - 15.7 \text{ m}^2/\text{s}^2$$

Here the displacement is up rather than down so we have made the 0.8 m negative. If this seems confusing, don't worry. For many people this seems weird (displacement up is one sign, displacement down, the other). Notice what would have happened if we did not make the 0.8 m negative, though. We would have ended up with the square root of a negative number. In the introductory physics course the answer is almost never the square root of a negative number, i.e. $\sqrt{-1} = i$, which is an imaginary number. It is good to keep in mind that if you ever do end up with the square root of a negative number, you need to go back and check your assignment of minus signs. Solving for v_{after}

 Remember

$$v_{after} = \sqrt{15.7 \text{ m}^2/\text{s}^2} = \pm 3.96 \text{m/s}$$

We have a choice of + or − for this value. First, mathematically, why did we stick a ± sign in front of the 3.96? Isn't it true that $(-3.96)^2 = 15.7$ just as easily as $+3.96^2 = 15.7$? There's nothing imaginary about **squaring** a negative number, only taking the **square root.** So here is another mathematical rule to remember, $\sqrt{4} = \pm 2$.

 Remember

Now, physically speaking, should we choose negative or positive for 3.96? Remember that in the problem we said that positive was down and negative was up, so $v_{after} = -3.96 \text{ m/s}$. The direction of this velocity is opposite from v_{before}, so it makes sense that the signs would be opposite. Now, since we are thinking about the problem as we go along, does the magnitude of this number make sense? Earlier we calculated $v_{before} = 5.34 \text{m/s}$ for the velocity of the ball when it was pushed downward from a height of 1 meter. We know that on the way back up the ball only rises to a height of 0.8 meters at which point it has no more velocity, so it does make sense that the magnitude of v_{after} is less than the magnitude of v_{before}.

The question asked for the impulse applied to the ball by the floor. We were going to use the impulse-momentum theorem:

$$\text{Impulse} = mv = mv_{after} - mv_{before}$$

$$\text{Impulse} = .045 \text{ kg}(-3.96 \text{ m/s}) - .045 \text{ kg}(5.34 \text{ m/s}) = -0.419 \text{ kg} \cdot \text{m/s}$$

The impulse of the floor on the ball is negative, which is consistent since we have always been calling the upward direction negative.

8-3 James Bond is skiing and is being pursued by Goldfinger, also on skis. Assume no friction. Mr. Bond, who has a mass of 100 kg, fires backward a 40g bullet at 800 m/s. Goldfinger, at 120 kg, fires forward at Bond with a similar weapon. What is the relative velocity change after the exchange of 6 shots each? No bullets hit Bond or Goldfinger (because they are both main characters and it is early in the movie).

Solution: The problem is analyzed with conservation of momentum. The momentum ($m_b v_b$ lets call it) of the bullet fired by Bond increases his momentum by $m_B \Delta v_B$. We are using little b to stand for the bullet and big B to stand for Bond. **The key here is to understand that whatever momentum the bullet has in one direction, Bond has in the exact opposite direction.** Remember that each of the 6 bullets Bond fires increases his velocity. Set $m_b v_b = m_B \Delta v_B$ and solve for Δv_B.

Insight

$$40 \times 10^{-3} \, \text{kg}(800 \, \text{m/s}) = (100 \, \text{kg})\Delta v_B \quad \text{or} \quad \Delta v_B = 0.32 \text{ m/s}$$

The $40 \times 10^{-3} \, \text{kg}$ bullet is small compared to the 100 kg of Bond and it would not be necessary to subtract that mass from Bond's mass in the calculation. The Δv_B notation is used to indicate that each bullet fired by Bond causes a change in his velocity.

Goldfinger, on the other hand, has his momentum decreased. In his case, $m_b v_b = m_G \Delta v_G$. Putting in the numbers

$$32 \, \text{kg} \cdot \text{m/s} = (120 \, \text{kg})\Delta v_G \quad \text{or} \quad \Delta v_G = 0.26 \text{ m/s}$$

Bond goes faster and Goldfinger goes slower with the total change in velocity 0.58m/s for each pair of shots fired. For six shots, this amounts to a difference of 3.48m/s. If Bond and Goldfinger had been travelling at the same speeds, then after this exchange Bond would have a relative speed advantage of 3.48 m/s with which to make his escape.

Keep in mind this application of conservation of momentum when doing rocket propulsion problems. The gases expelled from a rocket have a certain momentum in one direction, and the rocket has an equivalent change in momentum in the opposite direction.

8-4 A ballistic pendulum, a device for measuring the speed of a bullet, consists of a block of wood suspended by cords. When the bullet is fired into the block, the block is free to rise. How high does a 5.0 kg block rise when a 12 g bullet travelling at 350 m/s is fired into it?

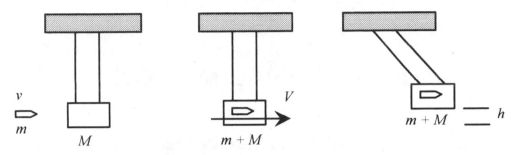

Solution: This is a classic physics problem that many instructors like to put on tests. The collision between the bullet and the block is **inelastic** (the bullet comes to rest in the block). Part of the kinetic energy of the bullet **goesinto** friction as the bullet burrows its way into the block. Therefore kinetic energy is not conserved. Momentum is conserved, however, because momentum is always conserved in collisions.

Because the collision is inelastic, apply conservation of momentum to the bullet-block collision. Before the collision all the momentum is in the mv of the bullet.   After the collision the momentum is in the $(m+M)V$ of the block and bullet. We assume that the bullet comes to rest (transfers all its momentum) before there is appreciable motion of the bullet-block combination.

$$mv = (m+M)V$$

After the collision, the rise of the block is determined by energy analysis. The kinetic energy of the block **goesinto** potential energy.

$$(1/2)(m+M)V^2 = (m+M)gh \quad \text{or} \quad V^2/2 = gh$$

Substituting for V from $mv = (m+M)V$, we get

$$\frac{1}{2}\left(\frac{m}{m+M}\right)^2 v^2 = gh \quad \text{so} \quad v = \frac{m+M}{m}\sqrt{2gh} \quad \text{or} \quad h = \frac{v^2}{2g}\left(\frac{m}{m+M}\right)^2$$

giving the relation between the velocity of the bullet and the height the block and bullet rise. For this problem

$$h = \frac{(350\,\text{m/s})^2}{2 \cdot 9.8\,\text{m/s}^2}\left(\frac{0.012}{5.012}\right)^2 = 3.6 \text{ cm}$$

In this problem, the 0.012 can be neglected in comparison to 5.0. This is not always the case so we write $m+M$ as 5.012 as a reminder to include both $m+M$ in the calculation. **Remember**

9

ROTATIONAL MOTION

Things to Remember

- The definition of angle (θ) is arc length (s) over radius (r). This gives the angle as a pure number, or radians. Radians is a phantom unit. Sometimes it is used and sometimes it is not. For an arc length of $4\,\mathrm{m}$ on a circle of radius $2\,\mathrm{m}$, the angle in radians is $4\,\mathrm{m}/2\,\mathrm{m}=2$. This is usually written as $2\,\mathrm{rad}$. In a problem involving canceling units this can add confusion because rad is length over length or unity and does not cancel with anything.

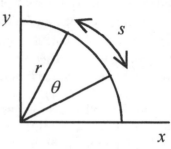

So rad may appear in a problem, then disappear in the answer, or vice versa, thus fitting the description "phantom." When radians is used in a problem the unit is for clarity and will not necessarily be in the final answer.

- The average angular speed is defined as $\overline{\omega} = \dfrac{\Delta\theta}{\Delta t} = \dfrac{\theta_2 - \theta_1}{t_2 - t_1}$. The units are rad/sec.

  This is analogous to the linear definition of average velocity, which is $\overline{v} = \dfrac{\Delta x}{\Delta t}$. By the way, ω is the Greek letter omega (not to be confused with the English letter w). The line above the omega signifies average angular velocity as opposed to instantaneous angular velocity.

- The average angular acceleration is defined as $\overline{\alpha} = \dfrac{\Delta\omega}{\Delta t}$. The units are rad/sec^2. This is analogous to linear acceleration, which is $\overline{a} = \dfrac{\Delta v}{\Delta t}$.

- There is a set of equations for rotational motion that are analogs of the equations that we learned earlier for linear motion with constant acceleration. These should be memorized or placed on the sheet of paper that you have for important equations to remember.

<u>Constant linear acceleration</u>   <u>Constant angular acceleration</u>

$$v = v_0 + at$$

$$x = \frac{v_0 + v}{2}t$$

$$x = v_0 t + (1/2)at^2$$

$$v^2 = v_0^2 + 2ax$$

$$\omega = \omega_0 + \alpha t$$

$$\theta = \frac{\omega_0 + \omega}{2}t$$

$$\theta = \omega_0 t + (1/2)\alpha t^2$$

$$\omega^2 = \omega_0^2 + 2\alpha\theta$$

- Another set of relationships in rotational motion is the one between linear motion and angular motion. These relate the motion of a particle, or point, along a circular path to the angular motion. Position along the circular path is called the linear position or rim position. The first equation is $\Delta s/\Delta t = r(\Delta\theta/\Delta t)$ or $v = r\omega$. This relates the linear velocity, the velocity of a point on the rim, to the angular velocity.

- Also there is $\Delta v/\Delta t = r(\Delta\omega/\Delta t)$ or $a = r\alpha$ relating the linear acceleration, the acceleration of a point on the rim, to the angular acceleration.

9-1 What is the angular velocity of the second hand on a clock in radians/s?

Solution: The second hand completes one revolution in 60 s, so we need to convert from revolutions to radians.

$$\frac{1 \text{ revolution}}{60 \text{ s}} \times \frac{2\pi \text{ radians}}{\text{revolution}} = 0.105 \text{ radians/s}$$

This conversion comes from the fact that the circumference of a circle is $2\pi r$. So $2\pi r = 1$ revolution = 360°. In general you will want to switch revolutions to either radians or degrees depending on which is more convenient.

★ **Remember**

9-2 A wheel accelerates uniformly from rest to an angular velocity of 25 rad/s in 10 revolutions. What is the angular acceleration of the wheel?

Solution: Let's solve this the same way we solved similar problems with linear acceleration in chapter 2. Write down the 4 equations of motion and write down what we know and look for the equation that fits.

 Pattern

<u>Equations</u>   <u>Things we know</u>

$$\omega = \omega_0 + \alpha t$$ $\omega_0 = 0$ (starts from rest)

$$\theta = \frac{\omega_0 + \omega}{2}t \qquad\qquad \omega_f = 25\,\text{rad/s}$$

$$\theta = \omega_0 t + (1/2)\alpha t^2 \qquad \theta = 10 \text{ revolutions}$$

$$\omega^2 = \omega_0^2 + 2\alpha\theta \qquad\qquad \alpha = ?$$

First, notice that we have ω_f in units of rad/s and we have θ in terms of revolutions. First, convert revolutions to radians.

$$\theta = 10 \text{ revolutions} \times \frac{2\pi \text{ rad}}{1 \text{ rev}} = 62.8 \text{ rad}$$

Now find the equation that we can use. The first three equations all have t in them and t is not given in the problem. The last equation has everything that we need.

$$\omega^2 = \omega_0^2 + 2\alpha\theta \implies (25 \text{ rad/s})^2 = 0^2 + 2\alpha(62.8 \text{ rad})$$

$$\alpha = 4.98 \text{ rad/s}^2$$

And only a nerd wouldn't just call this 5 rad/s^2.

9-3 An airplane engine is started up. The propeller accelerates from rest with an angular acceleration of 4.0 rad/s^2 for one minute. The engine then stalls and the propeller decelerates until it comes to rest again two minutes later. How many revolutions did the propeller make?

Solution: Let's solve this problem in two parts. First we will find how many revolutions there are when the propeller is accelerating, then move on to the decelerating portion of the problem.

Equations   Things we know

$$\omega = \omega_0 + \alpha t \qquad\qquad \omega_0 = 0 \text{ (starts from rest)}$$

$$\theta = \frac{\omega_0 + \omega}{2}t \qquad\qquad t = 60\text{s}$$

$$\theta = \omega_0 t + (1/2)\alpha t^2 \qquad \alpha = 4.0 \text{ rad/s}^2$$

$$\omega^2 = \omega_0^2 + 2\alpha\theta \qquad\qquad \theta = ?$$

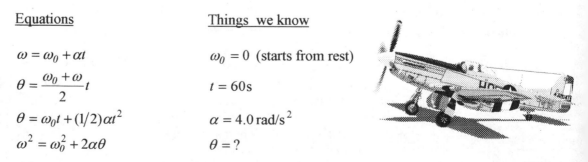

For the first part of the problem, we use the third equation, $\theta = \omega_0 t + (1/2)\alpha t^2$. Since we are breaking the problem up into two parts, let's add the subscript 1 to each quantity. That way we won't end up confusing values calculated in the first part of the problem with values calculated in the second part of the problem. So therefore the equation becomes

$$\theta_1 = \omega_{01} t_1 + (1/2)\alpha_1 t_1^2$$

$$\theta_1 = 0 + (1/2)(4.0 \text{ rad}/\text{s}^2)(60 \text{ s})^2 = 7.2 \times 10^3 \text{ rad} \times \frac{1 \text{ rev}}{2\pi \text{ rad}} = 1.15 \times 10^3 \text{ rev}$$

Now for the second part of the problem, we know that the final velocity is 0 ($\omega_2 = 0$) and that $t_2 = 120$ s. Unfortunately, these two pieces of information are not enough to solve for θ_2 in any of the equations. We could use $\theta_2 = \frac{\omega_{02} + \omega_2}{2} t_2$ if we knew ω_{02} , the starting velocity for the second half of the question. Think on your own now, is there another means of getting this value?

The answer is, yes there is. The value ω_{02}, the starting velocity for the second half of the problem, is also the final velocity, ω_1, for the first half of the problem. Looking back at the information specified for the first part of the problem, do you see that we could use $\omega_1 = \omega_{01} + \omega_1 t_1$ to solve for ω_1?

$$\omega_1 = 0 + 4.0 \text{ rad/s}^2 (60 \text{ s}) = 2.4 \times 10^2 \text{ rad/s}^2 = \omega_{02}$$

$$\theta_2 = \frac{2.4 \times 10^2 \text{ rad/s}^2 + 0}{2} (120 \text{ s}) = 1.44 \times 10^4 \text{ rad} \frac{1 \text{ rev}}{2 \text{ rad}} = 2.29 \times 10^3 \text{ rev}$$

The answer to the question of how many revolutions the propeller goes through is obtained by adding $\theta_1 + \theta_2$: $1.15 \times 10^3 \text{ rev} + 2.29 \times 10^3 \text{ rev} = 3.44 \times 10^3 \text{ rev}$ or   3.44 thousand revolutions.

The pattern seen in this problem comes up in many of the harder textbook problems. Just to simplify it, suppose the following:

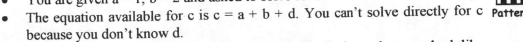

- You are given a = 1, b = 2 and asked to solve for c.
- The equation available for c is c = a + b + d. You can't solve directly for c because you don't know d.
- But if you consider other equations, you'll usually find one that may look like d = a/b. The way to get the answer is first solve for d, then plug that into c = a + b + d to get the answer for c.

Pattern

9-4 A compact disk (CD) operates such that the music is played back at a constant tangential speed at any point radially along the disk. Suppose that the CD rotates at 4 rad/sec at the outer edge, which is 5 cm from the center of the disk. What is the angular speed of the CD at the inner edge (2 cm from the center)?

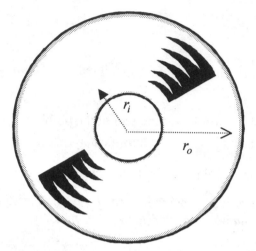

Solution: The equation that relates tangential velocity (or speed) to angular velocity is $v = r\omega$. And we need to have the same v at both the inner and outer radius. First, find v using the outer radius and outer angular velocity (ω_0, r_o).

$$v = \frac{4 \text{ rad}}{\text{sec}} \times \frac{5 \text{ cm}}{} = 20 \text{ cm/s}$$

We need this same v at the inner radius, so use the same equation to solve for ω_i.

$$20 \text{ cm/s} = \omega_i (2 \text{ cm}) \implies \omega_i = 10 \text{ rad/s}$$

10

ROTATIONAL DYNAMICS

Things to Remember

- The general definition of moment of inertia is $I = \sum mr^2$. The symbol $\sum$ stands for sum. The best way to understand this symbol is to work through problem 10-1. Moment of inertia is to rotational motion as mass is to translational (straight line) motion. Check your text for a table of moments of inertia for common shapes such as spheres, cylinders, and hoops.

- The definition of torque is $\sum \tau = I\alpha$. Torque is to rotational motion as force is to translational motion. Compare the equation for torque to $F = ma$. Remember that from chapter 5 we also know that $\text{torque} = \text{force} \times \text{lever arm}$. Review chapter 5 if you don't remember this.

- Rotational kinetic energy is defined as $(1/2)I\omega^2$. This should look similar to translational kinetic energy, which is $(1/2)mv^2$.

- Angular momentum is defined as $L = I\omega$. This is analogous to linear momentum, which is $p = mv$. Also, to go from linear momentum to angular momentum, use $L = mvr$.

- The law of conservation of angular momentum states that angular momentum in a system is conserved (an object that is spinning stays spinning), unless acted on by an outside torque. This is analogous to the law of conservation of linear momentum, which states that an object in straight-line motion stays in motion unless acted on by an outside force.

Moment of Inertia

Suppose that you have a basketball and a sphere of lead, both with the same radius. Which one is harder to spin at a given angular velocity? The lead ball, of course. The lead ball has the larger mass and therefore the larger moment of inertia.

Now suppose that you have a 1 pound ball on the end of a string. The ball is attached to a post. You are rotating the ball in a horizontal circle. Does it take more torque (force) to rotate the ball in a 1 meter radius (on the right below), or a 5 meter radius with the same angular velocity (say 1 revolution every 10 seconds)?

The answer is that it takes more torque to rotate the same mass in a big circle than in a little circle. Therefore moment of inertia about an axis is also greater for larger radius. The first type of problem with moment of inertia is one where there are several independent masses rotating around an axis and you have to add them all up to find the total moment of inertia.

10-1 Three masses are held together by light bars (weightless). Find the moment of inertia about the y axis as indicated by the dotted line.

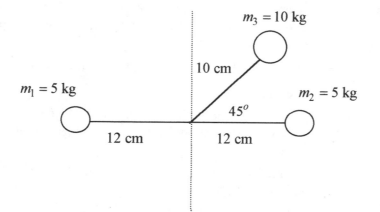

Solution: We need to use $I = \sum mr^2$. Another way of writing this is $I = m_1 r_1^2 + m_2 r_2^2 + m_3 r_3^2$. Now,

$$m_1 r_1^2 = 5 \text{ kg} (12 \text{ cm})^2 = 720 \text{ kg} \cdot \text{cm}^2$$

The units of moment of inertia are $\text{mass} \times \text{length}^2$. It may seem strange to label the units this way, but please remember to keep track of the units you are using in a problem. The value for $m_2 r_2^2$ is the same since it is the same mass at the same distance from the axis of rotation.

The tricky part here is with $m_3 r_3^2$. We are not talking about rotating about a point at the center of the masses; we are talking about rotating about an **axis** shown by the dotted line. Therefore r_3 is not 10 cm. Instead, it is the distance of the 10 kg mass from the y axis, which is $10 \text{ cm} \times \cos 45^\circ = 7.07 \text{ cm}$. The answer to the question, "what is I?", is:

Watch Out!

$$I = 2 \times 720 \text{ kg} \cdot \text{cm}^2 + 10 \text{ kg}(7.07 \text{ cm})^2 = 1940 \text{ kg} \cdot \text{cm}^2$$

Rotational Dynamics

10-2 A hoop of mass 1.0 kg and radius 0.25 m is rotating in a horizontal plane with angular momentum $4.0 \text{ kg} \cdot \text{m}^2/\text{s}$. A lump of clay of mass 0.20 kg is placed (gently) on the hoop. What happens to the angular velocity of the hoop?

Solution: The first thing that should enter your mind is "conservation of angular momentum". Something is spinning and we changed the mass, but applied no torque (or force) to slow it down. The angular momentum is conserved and the object spins more slowly. Try to picture this. The hoop is spinning. It has a certain mass and a certain amount of angular momentum. If the mass is suddenly increased without disturbing the system, that same amount of angular momentum now has more mass to spin. The hoop therefore slows down (angular velocity decreases).

Insight

First we need to find I, the moment of inertia. Find the area in your textbook where the moment of inertia of a hoop is given. There should be a table that has the moment of inertia of a sphere, cylinder, hoop, etc. The moment of inertia for a hoop is $I = mr^2$.

$$I = 1.0 \text{ kg}(0.25 \text{ m})^2 = 0.0625 \text{ kg} \cdot \text{m}^2$$

The angular momentum L is $I\omega$, so the initial angular velocity can be calculated from $L = I\omega_i$

$$4.0 \text{ kg} \cdot \text{m}^2/\text{s} = (0.0625 \text{ kg} \cdot \text{m}^2)\omega_i \quad \text{or} \quad \omega_i = 64 \text{ rad/s}$$

Since the lump of clay is placed (gently) on the hoop (the hoop is rotating in the horizontal plane and the clay is placed on the hoop vertically), placing the clay only adds mass to the hoop. This additional mass adds "another" moment of inertia, mr^2, but the **angular momentum for the system remains constant**. Therefore, we set this same angular momentum equal to the sum of the two moments of inertia and find the new angular velocity:

$$4.0\,\text{kg}\cdot\text{m}^2/\text{s} = \left[1.0\,\text{kg}(0.25\,\text{m})^2 + 0.2\,\text{kg}(0.25\,\text{m})^2\right]\omega_f \quad\text{or}\quad \omega_f = 53.3\,\text{rad/s}$$

10-3 An amusement park game consists of a paddle wheel arrangement where you shoot at the paddles with a pellet gun, thereby turning the wheel. The paddle wheel is set in motion with an initial angular momentum of $200\,\text{kg}\cdot\text{m}^2/\text{s}$ and angular frequency of $4.0\,\text{rad/s}$. Suppose that you shoot eight 40g pellets at a speed of $200\,\text{m/s}$ at the paddles. The pellets hit the paddles at $0.80\,\text{m}$ radius, stick, and impart all their momentum to the wheel. Find the new angular momentum and the new angular frequency.

Solution: First calculate the initial moment of inertia of the paddle wheel from $L = I\omega$.

$$200\,\text{kg}\cdot\text{m}^2/\text{s} = I(4.0\,\text{rad/s}) \quad\text{or}\quad I = 50\,\text{kg}\cdot\text{m}^2$$

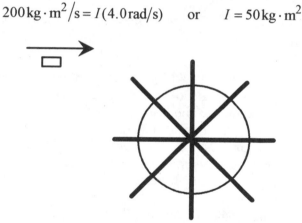

Now calculate the ω after the wheel has absorbed the momentum in the eight pellets. The angular momentum of each pellet is the linear momentum times the $0.80\,\text{m}$ radius.

$$L_{each\,pellet} = mvr = 0.040\,\text{kg}(200\,\text{m/s})0.80\,\text{m} = 6.4\,\text{kg}\cdot\text{m}^2/\text{s}$$

For eight pellets the "absorbed" angular momentum is $L_{all\,pellets} = 51.2\,\text{kg}\cdot\text{m}^2/\text{s}$.
The new angular momentum is the original plus this, or $251.2\,\text{kg}\cdot\text{m}^2/\text{s}$.

The I also has increased because of the additional mass at the $0.80\,\text{m}$ radius. This "additional" I is

$$I_{pellets} = 8 \times 0.040\,\text{kg}(0.80\,\text{m})^2 = 0.205\,\text{kg}\cdot\text{m}^2$$

The equation for calculating the new angular velocity ($L = I\omega$) is

$$251.2\,\text{kg}\cdot\text{m}^2/\text{s} = (50.2\,\text{kg}\cdot\text{m}^2)\omega_f \quad \text{or} \quad \omega_f = 5.00\,\text{rad/s}$$

10-4 Now consider a 10 kg solid cylindrical drum with radius 1.0 m rotating about its cylindrical axis under the influence of a force produced by a 30 kg mass attached to a cord wound around the drum. What is the torque, moment of inertia, and angular acceleration?

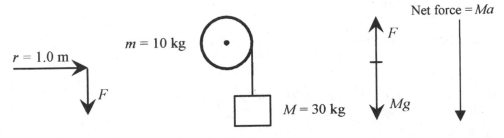

Solution: Moment of inertia is the easiest one to get first. The moment of inertia for a solid drum rotating about its axis is $mr^2/2$. This formula should be in the table in your text.

$$I = mr^2/2 = 10\,\text{kg}(1.0\,\text{m})^2/2 = 5.0\,\text{kg}\cdot\text{m}^2$$

Now, to the torque. The torque on the drum is the tension in the cable times the radius of the drum (torque = force × lever arm). Let's get force. The system accelerates clockwise so the unbalanced force on M makes it accelerate according to

$$Mg - F = Ma \quad \text{or} \quad F = M(g-a)$$

To see why this equation is true, look at the force diagrams on the far right above. There is the force of gravity, Mg, acting down. Opposing that is F, the tension in the cable. The block is falling however, so we do know that there is a net Ma downward. If there were no string attached to the block then Ma would just be equal to Mg. If we were constantly pushing on the block downward as it fell, then Ma would be Mg plus that pushing force. In this case, the block is not falling as fast as it normally would under the force of gravity because there is a cable attached and an upward force reducing its acceleration. The total net Ma on the falling block is therefore gravity minus the force from the cable.

We have the force, now we need to get torque. Unfortunately, we don't know a, the net linear acceleration for the system. So what do we do when we have two unknowns in an equation? Start generating more equations with those same unknowns, right?

Let's use both definitions for torque (torque = force × lever arm and $\tau = I\alpha$) and set them equal to each other: $rF = I\alpha$.

The other unknown we are looking for, α, is related to the tangential acceleration by $a = r\alpha$. Now we need to do a little algebra.

$$rF = \frac{mr^2}{2} \times \frac{a}{r} \quad \text{or} \quad F = \frac{ma}{2} \qquad \text{leading to} \qquad \frac{ma}{2} = Mg - Ma \quad \text{or} \quad a = \frac{g}{1 + m/2M}$$

Evaluating $a = \dfrac{9.8\,\text{m/s}^2}{1 + 10/60} = 8.4\,\text{m/s}^2$

The angular acceleration is $\alpha = \dfrac{a}{r} = \dfrac{8.4\,\text{m/s}^2}{1.0\,\text{m}} = 8.4\,\text{rad/s}^2$

The torque is $\tau = I\alpha = (5.0\,\text{kg} \cdot \text{m}^2)8.4\,\text{rad/s}^2 = 42\,\text{N} \cdot \text{m}$

10-5 When the 30 kg mass in problem 10-4 has fallen (starting from rest) through 4.0 m, what is the linear velocity, angular velocity of the drum, and the time for this to occur?

Solution: This part of the problem is approached from an energy point of view. The 30 kg mass falls through a distance of 4.0 m. The work performed by gravity **goesinto** translational *KE* of the 30 kg mass and <u>rotational</u> KE of the cylinder. <u>The main difficulty in problems like this is remembering that some of the energy results in rotational KE and some in translational KE.</u>

Insight

Gravitational potential energy = translational KE + rotational KE

$$Mgh = \frac{1}{2}Mv^2 + \frac{1}{2}I\omega^2 = \frac{1}{2}Mv^2 + \frac{1}{2}\frac{mr^2}{2}\frac{v^2}{r^2} \implies Mgh = v^2\left(\frac{M}{2} + \frac{m}{4}\right)$$

The key step in this problem is to write I in terms of m, and ω in terms of v and r.

$$30\,\text{kg}(9.8\,\text{m/s}^2)(4.0\,\text{m}) = v^2\left(\frac{30\,\text{kg}}{2} + \frac{10\,\text{kg}}{4}\right) \quad \text{or} \quad v^2 = 67.2\frac{\text{m}^2}{\text{s}^2} \quad \text{or} \quad v = 8.20\frac{\text{m}}{\text{s}}$$

From $v = r\omega$ the angular velocity is $\omega = \dfrac{v}{r} = \dfrac{8.20\,\text{m/s}}{1.0\,\text{m}} = 8.20\dfrac{\text{rad}}{\text{s}}$

The time for this system to reach this velocity is from $\omega = \omega_0 + \alpha t$

$$t = \frac{\omega}{\alpha} = \frac{8.20\,(\text{rad/s})}{8.4(\text{rad/s}^2)} = 0.976\,\text{s}$$

11

FLUIDS

Things to Remember

- Density is mass per unit volume, $\rho = m/V$. The density of water is 1gram/cm^3. 1cm^3 is also the same as 1 milliliter (ml).

- Pressure is force per unit area, $P = F/A$. The unit most commonly used for pressure is the Pascal (Pa), which is Newtons/meters^2. Atmospheric pressure on the earth, or one atmosphere as it is called, is equal to $1.01 \times 10^5 \text{ Pa}$.

- The pressure under a certain height of water is the atmospheric pressure plus the quantity ρgh. You can remember ρgh as being analogous to mgh for gravitational potential energy.

- Archimedes' Principle: The buoyant force on an object that is either floating or under water is equal to the weight of the water displaced by the object.

- The "equation of continuity" or "mass flow equation" for liquid flowing through a pipe that changes cross sectional area is $A_1 v_1 = A_2 v_2$. A is the cross sectional area and v is the **velocity** of the liquid. Be careful in this chapter not to confuse little v, velocity, with big V, volume.

- Bernoulli's equation: $\text{Pressure} + (1/2)\rho v^2 + \rho gh = \text{constant}$. Sometimes ρgh is written as ρgy. It's the same thing, h (height) or y (y axis) is a vertical distance.

11-1 Convert 3.45 atmospheres of pressure to the units of lbs/inch^2.

Solution: When you encounter chapters that introduce a whole bunch of new units, like this chapter, it is to your advantage to know how to make these kinds of conversions.

$$\frac{3.45 \text{ atm}}{} \frac{1.10 \times 10^5 \text{ N}}{1 \text{ atm} \cdot \text{m}^2} \frac{1 \text{ lb}}{4.45 \text{ N}} \frac{1 \text{ m}}{100 \text{ cm}} \frac{1 \text{ m}}{100 \text{ cm}} \frac{2.54 \text{ cm}}{\text{inch}} \frac{2.54 \text{ cm}}{\text{inch}} = 55.1 \text{ lbs/inch}^2$$

Remember that there are 100 cm in 1 meter, $100 \times 100 \text{ cm}^2$ in 1 m^2, and 100^3 cm^3 in 1 m^3.

11-2 Water must be pumped 300 meters high to the top floor of a skyscraper. What would be the minimum gauge pressure at the ground level for a pipe leading all the way up?

Solution: First of all, what do we mean by **gauge pressure**? Gauge pressure ($P - P_a$) is the pressure above atmospheric pressure. For example, if you put a pressure gauge on a tire, you may measure 35 psi, or pounds per square inch. That's the gauge pressure. You have to remember that the atmosphere is pushing in the opposite direction on the tire with a pressure of 1 atmosphere or 14.7 psi. The absolute pressure (P) inside the tire is therefore 35 psi + 14.7 psi. When the tire is flat, the gauge pressure reads 0 and the absolute pressure inside the tire is the same as the atmosphere, or 14.7 psi. The equation that we need here is

Insight

$$P = P_a + \rho g h \text{ or } P - P_a = Gauge \ Pressure = \rho g h$$

$$\rho g h = 1000 \text{ kg/m}^3 (9.8 \text{ m/s}^2) 300 \text{m} = 2.94 \times 10^6 \text{ Pa}$$

If we had asked how much energy does it take to raise a 10 kg ball of lead up to a height of 300 meters, you would use *mgh*, to find the increase in gravitational potential energy. So remember that pumping fluids is $\rho g h$ instead of *mgh*.

11-3 Suppose you are filling up a pool with a water hose. The pool is 5 feet deep, 10 feet wide and 20 feet long. The velocity of the water coming out of the hose is 10 feet/s and the hose has a diameter of 1 inch. How many hours does it take you to fill the pool?

Solution: First let's calculate the volume of the pool. It is $5 \times 10 \times 20$ feet or 1000 ft^3. Now let's get the cross sectional area of the hose. That is going to be $\pi r^2 = 3.14 (0.5 \text{ in})^2 = 0.785 \text{ in}^2$. Let's convert that over to ft^2 since everything else has ft for units:

$$\frac{0.785 \text{ in}^2 (1 \text{ ft})^2}{(12 \text{ in})^2} = 0.0055 \text{ ft}^2.$$

Notice that this cross sectional area times the velocity will give us volume/time, which is what we need to answer the question.

$$0.055\,\text{ft}^2 \times \frac{10\,\text{ft}}{\text{s}} = 0.055\,\text{ft}^3/\text{s}$$

Just think of a cylinder of water that emerges from the hose each second. It is 1 foot long and has a cross sectional area of $0.0055\,\text{ft}^2$. So we have calculated the volume of that cylinder, which is the volume of water that comes out of the hose each second.

Insight

How long will it take $1000\,\text{ft}^3$ to come out of the hose, then?

$$1000\,\text{ft}^3 \times \frac{1\,\text{s}}{0.055\,\text{ft}^3} = 18{,}182\,\text{s} \times \frac{1\,\text{min}}{60\,\text{s}} \times \frac{1\,\text{hr}}{60\,\text{min}} = 5.05\,\text{hours}$$

As you can see, this problem is solved mostly by looking at the units and making them come out right.

11-4 A rectangular-shaped, open-top steel barge is 10 m by 3.0 m and has sides 1.0 m high. The mass of the barge is 8.0×10^3 kg. How much mass can the barge hold if it can safely sink 0.75 m ?

Solution: Whenever you see a problem asking about things that are floating, remember Archimedes' principle: A body partially or completely immersed in a liquid is buoyed up by a force (F_B) equal to the weight of the liquid displaced. For this problem, the volume **Remember** of water displaced is length times width times allowed depth of the barge, $10\,\text{m} \times 3.0\,\text{m} \times 0.75\,\text{m} = 22.5\,\text{m}^3$. In other words, the volume of the barge that is submerged is the same as the volume of water that is displaced.

$$F_B = \rho g V = 1000\,\text{kg/m}^3\,(9.8\,\text{m/s}^2)22.5\,\text{m}^3 = 2.2 \times 10^5\,\text{N}$$

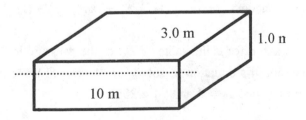

So this volume of water that is displaced has a weight, and this weight is the buoyant force. This buoyant force must support the barge (m_B) plus the load (m_L).

$$2.2 \times 10^5 \text{ N} = g(m_B + m_L), \quad m_B + m_L = 22 \times 10^3 \text{ kg}, \quad \text{and} \quad m_L = 14 \times 10^3 \text{ kg}$$

11-5 In Willy Wonka's chocolate factory, chocolate is flowing in a pipe as shown below. The chocolate has a density of 1.5 grams/cm^3. The flow is from left to right as indicated. As you can see from the drawing below, the flow is up a height of 20 meters and also into a narrower pipe. The velocity of the flow of chocolate and pressure in the pipe are known at the bottom, and the cross sectional area at the top is given. The factory workers (the umpa-lumpas) need to insure that the pressure in the pipe is great enough (in case anyone gets stuck in it and they need to be dislodged). Using the laws of fluid dynamics, calculate the pressure in the top section of the pipe.

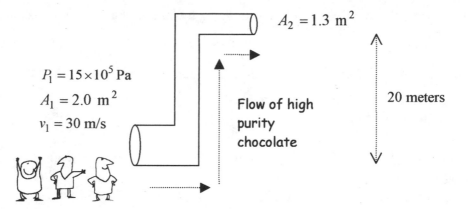

Solution: When you see a problem with fluids flowing in pipes, it is a good idea to write down the following two equations. The first is the "continuity equation" or "mass flow equation":

$$A_1 v_1 = A_2 v_2$$

Note that this term looks suspiciously like kinetic energy

The second is Bernoulli's equation:

$$P + (1/2)\rho v^2 + \rho g h = \text{constant} \quad \text{or} \quad P_1 + (1/2)\rho v_1^2 + \rho g h_1 = P_2 + (1/2)\rho v_2^2 + \rho g h_2$$

The calculation that is asked for is the value of P_2, or the pressure in the top section of the pipe. We can use Bernoulli's equation to solve for P_2, but first we need to get v_2 from the continuity equation. (Take $h_1 = 0$ and $h_2 = 20$ m.)

$$(2.0 \text{ m}^2)(30 \text{ m/s}) = 1.3 \text{ m}^2 (v_2) \implies v_2 = 46.2 \text{ m/s}$$

$$\frac{1.5 \text{g}}{\text{cm}^3} \times \frac{1 \text{kg}}{1000 \text{g}} \times \frac{(100 \text{cm})^3}{1 \text{m}} = 1500 \text{kg/m}^3$$

$$15 \times 10^5 \, \text{Pa} + (1/2)1500 \text{kg/m}^3 (30 \text{m/s})^2 + 0 =$$

$$P_2 + (1/2)1500 \text{kg/m}^3 (46.2 \text{m/s})^2 + 1500 \text{kg/m}^3 (9.8 \text{m/s}^2)(20 \text{m})$$

$$15 \times 10^5 \, \text{Pa} + 6.75 \times 10^5 \, \text{Pa} = P_2 + 16.0 \times 10^5 \, \text{Pa} + 2.94 \times 10^5 \, \text{Pa}$$

$$P_2 = 2.81 \times 10^5 \, \text{Pa}$$

Insight

This is a lower pressure than what we encountered in the larger part of the tube for two reasons:

1. When you go from a larger to a smaller cross sectional area, the pressure inside **drops**. This may not be intuitively obvious, but it is true. There are devices which use this principle to actually create suction. Next time you squirt water out of a garden hose, see if you can feel the difference in suction between a slow-moving stream and a narrower, fast-moving stream of water (without changing the amount of water flow from the faucet). Just feel the air near the stream of water.
2. The chocolate had to "climb up", so it did work against gravity and pressure was lost. This should seem intuitively correct. If not, then you are still utterly confused and need to review the problems on gravitational potential energy back in chapter 7.

11-6 Calculate the speed liquid flows out a small hole in the bottom of a large tank containing liquid to a depth of $1.0 \, \text{m}$.

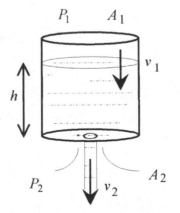

Solution: Apply the Bernoulli equation $P_1 + \rho g y_1 + \dfrac{1}{2}\rho v_1^2 = P_2 + \rho g y_2 + \dfrac{1}{2}\rho v_2^2$.

Note that $P_1 = P_2$, so remove these terms from the equation. Because $A_1 >> A_2$, v_1 is very small and is taken as zero. Take $y_1 = 0$ so y_2 is the depth of the liquid. The Bernoulli equation for this case reduces to

$$\rho g h = \frac{1}{2}\rho v_2^2 \quad \text{or} \quad v_2 = \sqrt{2gh}$$

The velocity of liquid from a small hole in a container depends only on the height of the column of liquid.

For a liquid height of 1.0 m $\quad v_2 = \sqrt{2(9.8\,\text{m/s}^2)1.0\,\text{m}} = 4.4\,\text{m/s}$

Insight Go back through this problem and note how the terms of the Bernoulli equation were handled, the pressure being the same, and the logic leading to the v_1 term being taken as zero. This is the hard part of this problem. If, in doing a problem like this, you do not see that v_1 can be taken as zero you will be stuck. You can move no farther on the problem because there is not enough data to calculate v_1.

12

HEAT AND CALORIMETRY

Heat - Things to Remember

- The conversion equation relating Fahrenheit and Celsius is $T_F = \dfrac{9}{5}T_C + 32$.

- The conversion from Celsius to Kelvin is $T_C = T_K - 273.15$.

- Remember the three equations for relating changes in length, area and volume of a substance to its change in temperature. They are: $\Delta L = \alpha L_0 \Delta T$, $\Delta A = \gamma A_0 \Delta T$, and $\Delta V = \beta V_0 \Delta T$.

- The ideal gas law: $PV = nRT$, where $R = 8.31\,\text{J/mol}\cdot\text{K}$. Just to be clear, the units are $\dfrac{\text{J}}{\text{mol}\cdot\text{K}}$, **not** $\dfrac{\text{J}}{\text{mol}}\cdot\text{K}$. Also, in order to get $PV = \text{Joules}$, you need to use Pascals and cubic meters. The ideal gas constant may also be expressed as $R = 0.0821\,\text{liter}\cdot\text{atm/mol}\cdot\text{K}$. Note the K in the bottom of the units for both values of R. That means you'd better use Kelvins for temperature whenever you use the ideal gas law.

- Avogadro's number is 6.02×10^{23} molecules/mole. Mole is like "dozen." You could have a dozen eggs (12 eggs/dozen), or a mole of eggs (6.02×10^{23} eggs/mole). You usually see eggs sold by the dozen, because eggs are big. If you're talking about oxygen atoms, then a dozen is not very many. You usually see quantities of molecules expressed in moles.

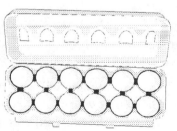

12-1 An aluminum pipe is installed when the temperature is $65^o\,\text{F}$. The pipe is 3.0 meters long. How long is the pipe when it is carrying steam at $300^o\,\text{F}$? The value for the linear expansion coefficient of aluminum is $\alpha = 24 \times 10^{-6}/^o\text{C}$.

Watch Out!

Solution: Don't put it past your physics instructor to give you a constant with units of Celsius and then ask a question where the temperature is given in Fahrenheit. In addition, you can bet that any trickster that would do that also has a multiple choice answer on the test that corresponds to what you get when you forget to make the conversion. Don't get bagged. Memorize (or write down on your equation sheet) the conversion formula. Let's get the temperatures over to Fahrenheit first.

$$T_F = \frac{9}{5}T_C + 32 \implies 65^o F = \frac{9}{5}T_C + 32 \implies \frac{5}{9}65 = T_C + 32\frac{5}{9} \implies T_C = 36.1 - 17.8 = 18.3$$

Just multiply every term in the
equation by the fraction 5/9.

And $300^o F \implies 148.8^o C$. The formula for the change in length with temperature is:

$$\Delta L = \alpha L_o \Delta T \implies \Delta L = 24 \times 10^{-6}/^o C \times 3.0 m \times (148.8^o C - 18.3^o C) = .0094 m$$

Insight

Remember that metals increase in size when they are heated up and decrease in size when cooled down. So the total length of the pipe is the initial length, L_o, plus the change in length, ΔL. The final answer is 3.0094 m.

12-2 A 5-liter tank is filled with nitrogen to a pressure of 40 atm. The temperature is $30^o C$. (a) What is the mass of the nitrogen inside the container? (b) How many molecules of nitrogen are in the container?

Solution: (a) The strategy here is to use the ideal gas law to find the number of moles of nitrogen and then convert moles to mass. But first, the inevitable units conversions. If you have $R = 8.31$ J/mol·K, then you need to remember that **only** if you use Pa for pressure and m^3 for volume, will you get Joules. However, if you have both expressions for R noted above at your fingertips for the test, then you won't have to waste time with units conversions. The other expression for R is 0.0821 liter·atm/mol·K.

Speed

$$PV = nRT \implies 40 atm\,(5 liters) = n(.0821\,liter \cdot atm/mol \cdot K)(273.15K + 30K)$$

$$n = 8.04\,moles \times \frac{28\;grams}{mol} = 225\;grams\,of\,nitrogen$$

The 28 grams/mol that suddenly appeared above is the molecular weight of nitrogen. There are two ways to get this number: 1. It may be given in a table in your text (or given on a test) or 2. You may need to figure it out from the periodic table of the elements.

Make it a point to find out at your next physics class whether you will be given this value or whether you will have to figure it out during a test.

If you have to figure it out, go now and look at the periodic table of the elements in your textbook. Notice that nitrogen, with symbol N, has the number 14 (or maybe 14.007) written in it's box. What that means is that the number of protons plus neutrons in a nitrogen **atom** is 14. The electrons are so small they have negligible mass compared to the protons and neutrons. But here's the snag – nitrogen is a diatomic molecule. That means that in the atmosphere, there are always two of them bonded together. So a mole of nitrogen means Avagadro's number of nitrogen **molecules**, not atoms. That's why we had to double the 14 to get to 28 grams/mol. It wouldn't be like that for the element helium. Notice it has an atomic weight of 4. Helium (He) is not diatomic, so if we had been doing the problem with He, we would have used 4 grams/mol.

OK, so how do you know if a gas is diatomic? Usually it is mentioned in the problem, but if not you can remember that the following gases are diatomic because they all rhyme: hydrogen, nitrogen, oxygen, and the halogens. (The halogens are the gases that make up group (column) 7 on the periodic table – fluorine, chlorine, etc.)

(b) The question also asked how many molecules were in the container. For this we use Avogadro's number:

$$n = 8.04 \text{ mol} \frac{6.02 \times 10^{23} \text{ molecules}}{\text{mol}} = 4.84 \times 10^{24} \text{ molecules}$$

Calorimetry

Remember

Heat is a form of energy. The unit of heat is the **calorie**, the heat to raise 1.0 g of water from $14.5^\circ C$ to $15.5^\circ C$. The kilocalorie or kcal is also popular. Food calories are actually kilocalories. The Joule equivalence is $1 \text{ cal} = 4.2 \text{ J}$.

The quantity of heat necessary to raise the temperature of a mass of material is $Q = mc\Delta T$ where c, the **specific heat**, is the amount of heat per $g \cdot ^\circ C$ for the material. The specific heat for water is $1.0 \text{ cal/g} \cdot ^\circ C$. Tables of specific heats are in most physics texts.

12-3 How much heat is required to raise a 1.8 kg copper teakettle containing 2.0 kg of water from $20^\circ C$ to $100^\circ C$? The specific heat of copper is $0.092 \text{ cal/g} \cdot ^\circ C$.

Solution: The total heat is $Q = \left[m_{cu} c_{cu} + m_{wat} c_{wat} \right] \Delta T$

$$Q = \left[1800\,g(0.092\,cal/g \cdot {}^{o}\,C) + 2000\,g(1.0\,cal/g \cdot {}^{o}\,C)\right]80^{o}\,C = 1.73 \times 10^{5}\,cal$$

The specific heat is sometimes expressed in J/kg·K rather than cal/g·^{o}C. For the previous problem the calculations would read

$$Q = \left[1.8\,kg(390\,J/kg \cdot K) + 2.0\,kg(4200\,J/kg \cdot K)\right]80\,K = 728,000\,J$$

Use the calorie-to-Joule equivalence to verify the numbers in this problem.

12-4 A 150 g cup of coffee (water) at 80°C has added to it 20 g of sugar (carbon) at 25°C. What is the final temperature of the insulated mixture? ($c_{carbon} = 0.12\,cal/g \cdot {}^{o}\,C$)

Pattern

Solution: In this problem, heat from the water raises the temperature of the carbon until they are both at the same temperature. This is a **goesinto** problem. We saw these types of problems earlier in the chapter on work and energy. We are also talking about heat energy flow in this problem, so the pattern is the same. The heat in the water **goesinto** heating the carbon until they both reach an equilibrium temperature, T. As a result, the water's temperature goes down, and the carbon's temperature goes up.

The heat leaving the water is $Q = mc\Delta T$ or $150\,g(1.0\,cal/g \cdot C^{o})(80 - T)^{o}\,C$.

The final temperature is between 80 and 25, so writing the temperature difference as $80 - T$ produces a positive number (of calories).

This heat goes to raise the temperature of the carbon $20\,g(0.12\,cal/g \cdot {}^{o}\,C)(T - 25)^{o}\,C$.

Notice that the temperature difference parenthesis are written so as to produce positive numbers.

The equation will read: The heat from the water **goesinto** heating the carbon.

$$150\,g(1.0\,cal/g \cdot C^{o})(80 - T)^{o}\,C = 20\,g(0.12\,cal/g \cdot C^{o})(T - 25)^{o}\,C$$

$$12,000 - 150T = 2.4T - 60 \quad \text{or} \quad T = 79.1^{o}\,C$$

Phase Change

If water is heated from room temperature to the vapor point it takes 1 calorie of heat (energy) to raise each gram 1 K degree. When the vapor point is reached a discrete amount of energy is required to convert each gram (or kilogram) of water at $100^o C$ to vapor (steam) **without raising its temperature** (this point is so important that it deserves both bold type and underlining). This amount of energy is called the **heat of vaporization** and for water it is $2.3 \times 10^6 J / kg$.

★ Remember

A similar phenomenon is observed when a solid is taken to the liquid phase. The **heat of fusion**, the heat to change $1.0 kg$ of ice to liquid water at $0^o C$ is $3.3 \times 10^5 J / kg$. Tables of heats of fusion and vaporization are in most physics texts. Problems involving heat of fusion or heat of vaporization must be done carefully because the energy requirements for these changes of state are very large. Converting a small amount of ice to water consumes a large amount of energy compared to raising the temperature of the water. It takes $4.2 \times 10^3 J$ per kg for each degree of temperature while it takes $3.3 \times 10^5 J$ per kg to convert ice to water with no temperature rise!

12-5 An insulated container holds $3.0 kg$ of water at $25^o C$. One kilogram of ice at $0^o C$ is added to the mixture. What is the final temperature and composition (ice and water) of the mixture?

Solution: First calculate the energy to bring all the water to $0^o C$.

$$Q = 3.0 \, kg \, (4200 \, J/kg \cdot K) 25 \, K = 315,000 \, J$$

This energy **goesinto** melting the ice. Now ask how much ice could be melted with this amount of energy.

$$315,000 \, J = (3.3 \times 10^5 \, J/kg) \, m \quad \text{or} \quad m = 0.95 kg$$

All the ice will not be melted so the final mixture will be $3.95 kg$ of water and $0.05 kg$ of ice all at $0^o C$. There will be no further heat transfer because both the water and the ice are at the same temperature.

12-6 If in the previous problem the amount of ice added to the water is reduced to $0.50 kg$, what is the final temperature of the composition?

Solution: Based on the previous calculation there is not enough ice to lower the temperature of the water to 0^oC. One way of doing the problem is to calculate the drop in temperature of the $3.0\,kg$ of water at 25^oC due to the melting ice, then treat the problem as a mixture of two amounts of water at different temperatures.

However, since we already know that all the ice melts and the final mixture is between 0^oC and 25^oC the equation for the final temperature can be set up with the statement: The energy in the 3.0 kg of water at 25^oC **goesinto** melting the ice and raising the temperature of the 0^oC water that results from the melted ice.

$$3.0\text{kg}\left(4200\,\text{J}/\text{kg}\cdot^o C\right)(25-T)^oC = 0.50\,\text{kg}\left(3.3\times10^5\,\text{J}/\text{kg}\right)+0.50\,\text{kg}\left(4200\,\text{J}/\text{kg}\cdot^o C\right)(T-0)^oC$$

$$315,000-12,600T = 165,000+2100T \qquad \text{or} \qquad T=10^oC$$

12-7 Two moles of helium are in a tank at 25^oC. Find (a) the total translational kinetic energy, (b) the kinetic energy per molecule, and (c) the rms speed of the atoms.

Solution: The purpose of this problem is to illustrate some of the types of calculations that are usually encountered in sections of your text called "kinetic theory of gases." (a) Pressure is caused by collisions of molecules with the walls of the container. The total translational kinetic energy of all the He molecules in the tank is:

$$E = \frac{3}{2}nRT = \frac{3}{2}2.0\,\text{mole}\frac{8.3\text{J}}{\text{mole}\cdot\text{K}}298\,\text{K} = 7420\,\text{J}$$

(b) The energy per molecule is: $\dfrac{m(v^2)_{avg}}{2} = \dfrac{3}{2}kT = \dfrac{3}{2}\left[\dfrac{1.4\times10^{-23}\text{J}}{\text{K}}\right]298\,\text{K} = 6.26\times10^{-21}\text{J}$

(c) The rms speed is (rms stands for **root mean square**)

$$v_{rms} = \sqrt{\frac{3RT}{M}} = \left[\frac{3(8.3\,\text{J/mol}\cdot\text{K})298\,\text{K}}{2.0\times10^{-3}\text{kg/mol}}\right]^{1/2} = 1930\frac{\text{m}}{\text{s}}$$

13

THERMODYNAMICS

Things to Remember

- The work done by an expanding gas is $W = P\Delta V$. If W is in Joules, then P is the pressure in Pascals and ΔV is the volume in cubic meters. Remember that in mechanics, Work was defined as $\text{Force} \times \text{Displacement}$. The equation for work in this chapter is analogous to that definition. Pressure is Newtons/m^2. If this is multiplied by volume, or m^3, then you end up with $\text{N} \cdot \text{m}$. These are the units of $\text{Force} \times \text{Displacement}$.

- Make a note now that if there is no volume change, then $\Delta V = 0$ and $W = 0$. Remember that in mechanics you can push as hard as you like, with as much force as you like on a wall, but if the wall doesn't move, you haven't done any work. The same is true with pressure. If you have a container of fixed volume you can crank up the pressure inside the container as high as you want, but if the volume of the container doesn't expand, there is no work done.

- The change in "internal energy" of a system is equal to the heat minus the work. Mathematically, $\Delta U = Q - W$.

- A gas that expands or is compressed **isothermally** has no change in temperature, and therefore, no change in internal energy.

- The efficiency of a Carnot engine is $Eff_C = 1 - \dfrac{T_C}{T_H}$.

  It is also true for heat flow that $Eff = 1 - \dfrac{Q_C}{Q_H}$.

- Entropy is a measure of the amount of disorder in a system. The formula in this chapter that relates to entropy is $\Delta S = \dfrac{\Delta Q}{T}$.

13-1 Three different experiments are run, in which a gas expands from point A to point D along the three paths shown below. Calculate the amount of work done for paths 1, 2, and 3.

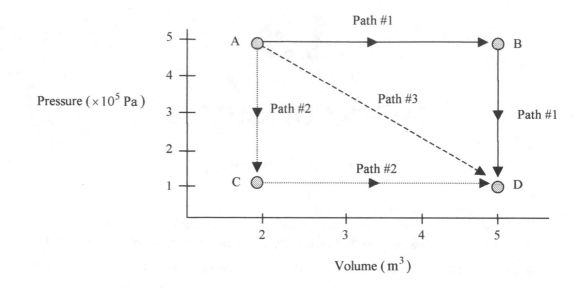

Solution: Path #1: The idea here is that we have gas in a container and we have independent control over both the pressure and the volume. So in path #1 we increase the volume from $2\,\text{m}^3$ to $5\,\text{m}^3$ without changing the pressure. Let's just pause for a moment and think about how we would realistically do that. One way would be to increase the temperature of the gas and have a flexible wall. If the temperature goes up, the pressure is going to go up if we keep the same volume, but what we are doing is constantly increasing the volume so that the pressure remains the same.

Remember that when gases are heated they expand. Heating a gas causes the molecules to have a higher velocity – that's what temperature is. If the molecules have a higher velocity, they hit the walls of a container with a higher velocity and cause a higher pressure – unless we expand the volume as we heat the gas in a way that keeps the pressure constant. All of these interactions are described by the ideal gas law $PV = nRT$.

Also, we are talking about **work** in this problem, not **heat** or temperature. Therefore, the fact that we would have to raise the temperature to maintain constant pressure while the volume increases is irrelevant to the question (and details like this are usually left out of these kinds of questions, making them hard to picture in your mind).

OK, back to the question. For path #1 the work done for the horizontal part is $W = P\Delta V = 5\times10^5\,\text{Pa}(3\,\text{m}^3) = 15\times10^5\,\text{J}$. The work done for the vertical part is 0 because there is no change in volume.

For path #2, the gas is lowered in pressure first at constant volume (perhaps the gas is cooled) and then the volume is expanded at constant pressure. Since the volume is expanded at a lower pressure, there is less work done. For path #2 we have $W = P\Delta V = 1\times10^5\,\text{Pa}(3\,\text{m}^3) = 3\times10^5\,\text{J}$. We got the same final conditions of temperature and pressure, but less work was involved. Remember, the question isn't telling us anything about how we got to these same conditions. Further, what happened to the **heat** in the gas is a different story.

Now it is time to talk about the area under the *PV* curve, because this is the fastest way to solve these problems. The areas under the *PV* curves for both path #1 and path #2 are shown below. Whether or not you understand all the details of the preceding paragraphs, just remember the bottom line: If you need to find the work done, take the area under the *PV* curve. Notice that the areas under these curves are the same as the answers we already obtained for path #1 and path #2 using physical principles.

Speed

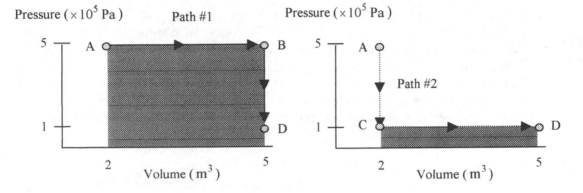

For path #3, we again take the area under the curve as shown. We need to include both the triangle and the rectangle beneath. The area of the triangle is ½ base×height or

$(1/2)3\,\text{m}^3 \times 4\times10^5\,\text{Pa} = 6\times10^5\,\text{J}$.

The rectangle underneath is $3\times10^5\,\text{J}$ (the same piece of the area that we found for path #2). So the total work done for path #3 is $9\times10^5\,\text{J}$.

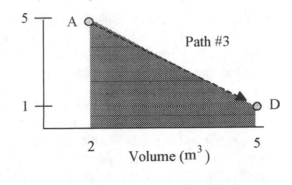

13-2 An ideal gas is compressed at a constant temperature of 50°C. 1.34×10^4 J of work is done on the gas. (a) What is the change in internal energy of the gas, and (b) what is the amount of heat that must be added or removed from the gas?

Solution: Most textbooks and instructors like to use a minus sign for work if work is done **on** the gas (i.e. the volume decreases), and a plus sign for work if work is done **by** the gas.

Remember

(a) The change in internal energy of the gas is 0. This part of the question is easy if you know the secret, but near impossible if you don't. So remember this: Internal energy depends on temperature. **If there is no temperature change in a gas, then there is no change in internal energy.** If you see the word **isothermal** used to describe a process, that means no change in temperature.

(b) To find the heat added or removed from the gas, we have $\Delta U = Q - W = 0$; $Q = W = -1.34 \times 10^4 \, \text{J}$. The negative number means that heat flows out of the gas.

Remember

Are you confused yet about the minus signs? We sure are! There's no real logic or consistency to the sign conventions in thermodynamics, so just memorize them. Don't bother trying to figure out why. As we said before, work done **on** a gas is negative (volume compression) and work done **by** a gas is positive (volume expansion). With heat it goes like this: heat added to the gas is positive, heat sucked out of the gas is negative. These are the sign conventions most commonly used. You may want to verify that this is how your textbook and instructor treat minus signs for work and heat in thermodynamics. Now would be a good time to check your textbook.

13-3 A Carnot engine operates between a heat reservoir of 179 Kelvins and a cold reservoir of 50 Kelvins. What is the maximum theoretical efficiency of this engine?

Solution: The Carnot cycle is an important theory to understand. Problems that include the analysis of the Carnot cycle appear frequently on tests. Please review your text and ask your instructor, teaching assistant, or classmate to help you understand this cycle. Here are some of the highlights of the Carnot cycle from the theoretical standpoint.

The basic idea is that you have a cylinder full of an **ideal gas**. On one side of the cylinder you have the heat reservoir, on the other side, a cold reservoir. All this means is that one side is hot and one side is cold, respectively. Heat flows from the hot reservoir to the cold reservoir and the engine provides mechanical work. Remember that we looked at three paths to go from point A to point D in problem 13-1? Remember that any consideration of what was happening to the temperature did not enter into the calculation? We just looked at the work done. The Carnot cycle looks at both work and temperature considerations. What this guy Carnot (who figured this stuff out in the early 1800's) did was find the most efficient way to get from point A to point D and then back to point A (referring back to the initial diagram in problem 13-1). By "the most efficient way," we mean that he found out how to get the most work out, given a certain amount of heat in. Remember this point: The Carnot engine trades heat for work. The trade isn't 100%

efficient though. You don't get 100 J of work for 100 J of heat. It's like taxes; there's an automatic deduction. Fortunately, calculating how much heat is deducted before you get work out is no where near as complicated as the IRS code. The equation is simple:

$$Eff_C = 1 - \frac{T_C}{T_H} = 1 - \frac{50\,K}{179\,K} = 1 - 0.28 = 0.72 = 72\%$$

This means that for 100 J of heat, you get 72 J of work. So where did the 28% of the heat that we lost go? A complete analysis of where this 28% goes is beyond the scope of this book, but we can say that in general, it increases the entropy of the universe.

13-4 A heat engine operates in a Carnot cycle between the temperatures of 100 and 400 Kelvins. The duration of each cycle is 1.5 seconds, and the engine absorbs 25 kJ of heat per cycle from the hot reservoir. (a) What is the maximum power output of this engine? (b) How much heat is expelled in each cycle?

Solution: (a) Power is energy per unit time. The strategy is first to find how many Joules of work is done in each cycle, and then divide that by 1.5 seconds to get the power. First let's calculate the efficiency:

$$Eff_C = 1 - \frac{T_C}{T_H} = 1 - \frac{100\,K}{400\,K} = 1 - 0.25 = 0.75 = 75\%$$

$$0.75 \times 25\,kJ = 18.8\,kJ \quad \text{the power then is:} \ 18.8\,kJ/1.5\,s = 12.5\ kW$$

(b) In order to find the amount of heat expelled per cycle, you need to realize that the following formula also works:

$$Eff = 1 - \frac{Q_C}{Q_H} \quad \Rightarrow \quad 0.75 = 1 - \frac{Q_C}{25\,kJ} \quad \Rightarrow \quad Q_C = 6.25\ kJ$$

Here Q_H refers to the heat flowing from the hot reservoir and Q_C refers to the heat expelled into the cold reservoir. The problem was nice in that it gave us temperature in Kelvins. Even in this equation for efficiency, you need to use Kelvin, not Celcius.

13-5 Two kilograms of ice at $0^o C$ melts to water at $0^o C$. What is the change in entropy?

Solution: The change in entropy is given by $\Delta S = \dfrac{\Delta Q}{T}$. First we need the change in heat, ΔQ. Remember that when a substance changes phase (liquid to solid or liquid to gas) without changing temperature, the amount of heat involved is determined by the heat of fusion (or heat of vaporization) for the substance. Here we are talking about going from solid to liquid, so we need the heat of fusion for water, which is 3.35×10^5 J/kg. Do the units of heat of fusion refresh your memory about the equation that relates heat in Joules to heat of fusion? Hint: What units do you need to multiply J/kg by to get J? The equation is:

$$\Delta Q = L_f m = (3.35 \times 10^5 \text{ J/kg}) \times 2 \text{ kg} = 6.7 \times 10^5 \text{ kg}$$

And plugging into the formula for entropy we get:

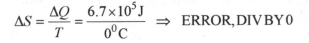

$$\Delta S = \frac{\Delta Q}{T} = \frac{6.7 \times 10^5 \text{ J}}{0^0 \text{C}} \;\Rightarrow\; \text{ERROR, DIV BY 0}$$

What's wrong here? Take a wild guess as to what the units of entropy are...........
If you guessed Joules/**Kelvin**, then congratulations, you are no longer **_utterly confused_**. The correct answer is

$$\Delta S = \frac{\Delta Q}{T} = \frac{6.7 \times 10^5 \text{ J}}{273.15 \text{ K}} = 2.45 \times 10^3 \text{ J/K}$$

14

SPRINGS, WAVES, AND SOUND

Springs

14-1 A 10 kg mass is hung vertically from a spring. The spring stretches 6 cm from its equilibrium position. What force is needed to stretch the spring 12 cm?

Solution: First we need to find the spring constant, k. The force on the spring initially is provided by gravity:

$$F = ma = 10 \text{ kg} (9.8 \text{ m/s}^2) = 98 \text{ N}$$

Now the spring constant is:

$$F = -kx \implies -98\text{N} = -k(6 \text{ cm}) \implies k = 16.3 \text{ N/cm}$$

Using this spring constant we can now go back and calculate the answer.

$$F = -kx \implies F = -16.3 \text{ N/cm} (12 \text{ cm}) = -196\text{N}$$

What's going on with the minus sign in front of the force? Remember that when you use Hooke's law, $F = -kx$, the force that's in the equation is **the force of the spring on the mass**. The diagram below shows the original situation. The force of gravity is pulling in the same direction as the displacement of the spring, but the force of the spring on the mass is pulling in the opposite direction of the displacement, x. The point of the minus sign is to indicate that a spring pulls against the direction of the displacement. It resists being stretched. Since the question asked what force would be needed to stretch the spring 12 cm, our answer is positive, and is 196 N.

Insight

Spring in equilibrium

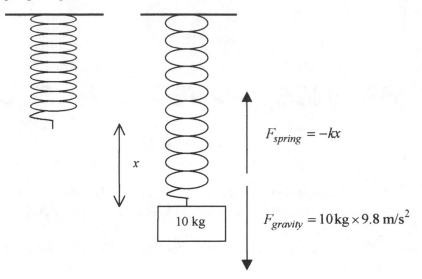

$$F_{spring} = -kx$$

$$F_{gravity} = 10\,kg \times 9.8\ m/s^2$$

14-2 A bullet with a mass of 15 grams is fired into a 20 kg block of wood which is attached to a spring as shown below. The velocity of the bullet is 350 m/s. If the spring constant is 30 N/m, how far does the spring compress? The bullet does not penetrate the block.

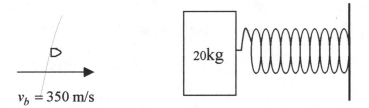

$$v_b = 350\ m/s$$

Solution: Does this problem look at all familiar? Do you remember the ballistic pendulum problem back in chapter 8? If not, go back and review problem 8-4.

The strategy for this problem is as follows:

1. First we consider the bullet striking the block where the initial collision is governed by **the law of conservation of momentum**.

2. Next we calculate the kinetic energy of the block + bullet immediately after impact (same as the ballistic pendulum problem so far). We then write the equation that describes the fact that all of the initial

Pattern

Blast from the past

kinetic energy of the block **goesinto** potential energy of the spring and calculate the compression of the spring.

1. The law of conservation of momentum for this situation is:

$$m_{bullet}v_{bullet} = (m_{bullet} + m_{block})v_{block}$$

$$0.015\,kg\,(350\,m/s) = (20.015kg)v_{block} \implies v_{block} = 0.262\,m/s$$

2. The kinetic energy of the block + bullet **goesinto** potential energy of the compressed spring and is calculated as follows:

$$(1/2)(m_{bullet+block})v_{block}^2 = (1/2)kx^2$$

$$(20.015\,kg)(0.262\,m/s)^2 = (30\,N/m)x^2 \implies x = 0.214\,m = 21.4\,cm$$

14-3 A spring hangs vertically with a 10 gram mass attached. It vibrates in simple harmonic motion with a frequency of 20 Hz. What is the spring constant?

Solution: First convert frequency to period: $f = 1/T$. The unit hertz (Hz) is the same as 1/s. So the period is (1/20) s or 0.05 s. Period is the amount of time for one complete oscillation. Frequency is just 1 divided by that time. Now we use the following formula:

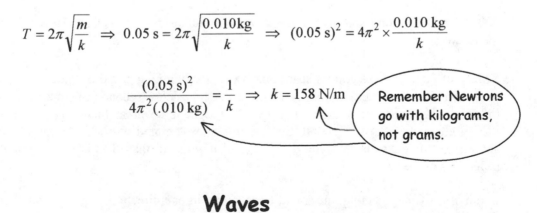

$$T = 2\pi\sqrt{\frac{m}{k}} \implies 0.05\,s = 2\pi\sqrt{\frac{0.010kg}{k}} \implies (0.05\,s)^2 = 4\pi^2 \times \frac{0.010\,kg}{k}$$

$$\frac{(0.05\,s)^2}{4\pi^2(.010\,kg)} = \frac{1}{k} \implies k = 158\,N/m$$

Remember Newtons go with kilograms, not grams.

Waves

14-4 A raft is floating on the surface of the ocean (it has zero speed on the surface of the water). As the waves pass, it rises and falls through a total distance of 20 m every 30 seconds. The crests of each wave are 30 m apart. (a) What is the amplitude of the wave? (b) What is the speed of the waves in the water?

Solution: (a) The amplitude is 10 m. Refer to the diagram and remember that amplitude is the distance between the middle point, or equilibrium, and maximum displacement, not the total distance between the maximum and minimum displacements.

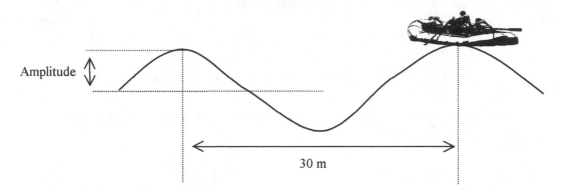

(b) To get the speed of the waves, we use *velocity = wavelength × frequency*. The wavelength is the distance between the crests, or 30m. The frequency is 1/the time between passing of the crests, or 1/(30 sec).

$$v = \frac{1}{30\ \text{sec}} 30\ \text{m} = 1\ \text{m/s}$$

Sound

14-5 A man shouts and hears his echo off a mountain 5 seconds later. How far away is the mountain?

Solution (a): Perhaps you've heard that if you see lightning and then wait for the sound of the thunder, that the lightning is 1 mile away for each 5 seconds of time between the lightning flash and the thunder. The speed of light is so great that it is instantaneous compared to the speed of sound. So you might be tempted to shout "1 mile!" for the answer to this question, but you'd be **bagged** if you did that. The correct answer is ½ mile. Do you see why?

Watch Out!

In the case of the lightning, the sound travels only in one direction. For the man shouting, the sound has to get to the mountain and then back.

Solution (b): The speed of sound in air at sea level (yes, it depends on height) is 345 m/s. (This is a number worth memorizing.)

Remember

$$345\ \text{m/s} \times 2.5\ \text{s} = 862.5\ \text{meters} = 0.862\ \text{km} \frac{1\ \text{mi}}{1.6\ \text{km}} = 0.54\ \text{mile}$$

14-6 Eight thousand fans at a football game scream "charge!!" at the prompting of the announcer. The decibel meter reads 80 dB for the sound level. How may fans would be needed to make the decibel meter read 90 dB, assuming they all scream with the same loudness?

Solution: If the answer were as easy as 9,000 then the question probably wouldn't have been asked. The real question here is, "What the heck is a dB?!" It stands for decibel. And decibels are weird. For starters, 0 dB is not no sound at all. It is the threshold of hearing (the softest whisper you can hear). Also

$1 dB + 1 dB \neq 2 dB$. What it comes down to is that every 10 dB increment is 10 times more sound intensity. So the answer to the question is that it would take 10 times as many fans, or 80,000 fans, to reach 90 dB. The dB scale is a logarithmic scale. An analogy would be that 10^9 is ten times larger than 10^8. Comparing the 9 and the 8 is analogous to comparing the 90 dB to the 80 dB.

14-7 Suppose you are cruising along at 70 miles per hour on the highway. An ambulance is approaching from the opposite direction with the siren on. At rest, the siren emits a frequency of 515 Hz. (a) What frequency do you hear as the ambulance approaches at 90 miles per hour? (b) What frequency do you hear as the ambulance recedes at 90 miles per hour? (c) The ambulance turns around and approaches you from the rear. Its

speed is 90 miles per hour, yours is 70. What frequency do you hear? (d) The ambulance passes you, now what frequency do you hear from behind?

Solution: This is a **Doppler shift** problem. The general equation is:

$$f' = f\left(\frac{v_{sound} \pm v_{observer}}{v_{sound} \mp v_{source}}\right)$$

(a) The tricky part about these Doppler shift problems is keeping the signs straight. Let's take the top part of the equation first. The positive sign is used for $v_{observer}$ if you are moving toward the source. Remember that the frequency (you hear frequency as "pitch") is increased if the observer moves toward the source. Now for the bottom part of the equation, notice that we used the $\mp$ symbol instead of the $\pm$. We know that if the source is moving toward the observer, the frequency has to go up also. In order for f' to go up, the denominator goes down. So we use the − sign in the denominator for v_{source} in the case where the source moves toward the observer. Converting the speeds of source and observer to m/s first,

$$\frac{90 \text{ mi}}{\text{hr}} \frac{1 \text{ hr}}{60 \text{ min}} \frac{1 \text{ min}}{60 \text{ s}} \frac{1.6 \text{ km}}{1 \text{ mi}} \frac{1000 \text{ m}}{\text{km}} = 40 \text{ m/s} \; ; \; 70 \text{ mi/hr} \Rightarrow 31.1 \text{ m/s}$$

$$f' = 515 \text{Hz} \left(\frac{345 \text{ m/s} + 40 \text{ m/s}}{345 \text{ m/s} - 31.1 \text{ m/s}} \right) = 632 \text{Hz}$$

(b) The signs are now reversed for this case:

$$f' = 515 \text{ Hz} \left(\frac{345 \text{ m/s} - 40 \text{ m/s}}{345 \text{ m/s} + 31.1 \text{ m/s}} \right) = 418 \text{ Hz}$$

(c) The car is moving away from the source, but the source is moving faster (gaining on) the car, so it should turn out that the frequency is a little higher. For this case we have:

$$f' = 515 \text{ Hz} \left(\frac{345 \text{ m/s} + 40 \text{ m/s}}{345 \text{ m/s} + 31.1 \text{ m/s}} \right) = 527 \text{ Hz}$$

(d) Now the source is receding and the car is following but not as fast as the source:

$$f' = 515 \text{ Hz} \left(\frac{345 \text{ m/s} - 40 \text{ m/s}}{345 \text{ m/s} - 31.1 \text{ m/s}} \right) = 500 \text{ Hz}$$

Insight

Probably the easiest overall strategy to have with Doppler shift problems is to know the one formula given in this problem. Then, if the velocity of the source is 0, just plug 0 into the equation. Also, if the observer is stationary and only the source is moving, just plug in 0 for the velocity of the observer.

15

ELECTRIC FORCES AND FIELDS

If this chapter is the beginning of the second semester of physics and you are starting here, guess what – you need to remember all that stuff about vectors from the first semester of physics. *Vectors* are going to make a big comeback in this chapter. So if you took the first semester of physics as a freshman and disliked it so much that it is now the last course you have to take in your senior year before you graduate, you need some review. If you are *utterly confused* about vectors, or if you've just forgotten everything you ever knew about them and then some, please review Chapter 1, Mathematical Background. We recommend that you follow along with your calculator and make sure that you are performing the trigonometric calculations correctly in this chapter. That's right, you need to remember all the trigonometry also.

New Things to Remember

- Like charges repel (two positive or two negative charges). Unlike charges have an attractive force between them.

- Coulomb's Law: The force between two charges is $F = k\dfrac{q_1 q_2}{r^2}$. k is $9 \times 10^9 \, \text{Nm}^2/\text{C}^2$. The C in that constant stands for Coulombs, which is a unit of charge. The charge of one electron is $1.6 \times 10^{-19} \text{C}$.

- The electric field due to a point charge is described by the formula $E = k\dfrac{q}{r^2}$. The units of electric field are either Newtons/Coulomb or Volts/meter. Memorize both units.

- The relationship between electric force and electric field is $E = F/q$.

- Begin to get into the habit of looking for **symmetry** in problems involving electric charge.

15-1   Two charges, $q_1 = 3.0 \times 10^{-8} \text{C}$ and $q_2 = -4.0 \times 10^{-8}\text{C}$, are separated by 6.0×10^{-3} m as shown below. What is the force of one on the other?

$$q_1 \qquad q_2$$

Solution: $\quad F = k\dfrac{q_1 q_2}{r^2} = 9.0 \times 10^9 \dfrac{\text{Nm}^2}{\text{C}^2} \dfrac{12 \times 10^{-16} \text{C}^2}{36 \times 10^{-6} \text{m}^2} = 0.30 \text{ N}$

This is the force each charge experiences due to the other. The fact that q_1 is positive and q_2 is negative means that the force is directed from one to the other along the line joining their centers.

15-2 Consider a line of charges, $q_1 = 8.0\,\mu\text{C}$ at the origin, $q_2 = -12\,\mu\text{C}$ at 2.0 cm, and $q_3 = 10\,\mu\text{C}$ at 4.0 cm as shown below. What is the force on q_3 due to the other two charges?

Solution: The force on q_3 due to q_1 is repulsive as shown. The magnitude of F_{31} is

$$F_{31} = 9.0 \times 10^9 \dfrac{\text{Nm}^2}{\text{C}^2} \dfrac{80 \times 10^{-12} \text{C}^2}{16 \times 10^{-4} \text{m}^2} = 450 \text{ N}$$

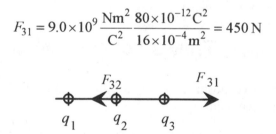

It is not necessary to be overly concerned about the sign of the force. The vector makes the direction clear. Drawing a vector diagram is a more sure way of getting the final sign correct than using algebraic signs in the force calculations.

The force on q_3 due to q_2 is attractive, and the magnitude is

$$F_{32} = 9.0 \times 10^9 \dfrac{\text{Nm}^2}{\text{C}^2} \dfrac{120 \times 10^{-12} \text{C}^2}{4.0 \times 10^{-4} \text{m}^2} = 2700 \text{ N}$$

The resultant force is $2700 \text{ N} - 450 \text{ N} = 2250 \text{ N}$. This force is attractive, and toward the left.

15-3 Three charges are arranged in the form of an equilateral triangle as shown below. Find the force on q_3 due to q_1 and q_2.

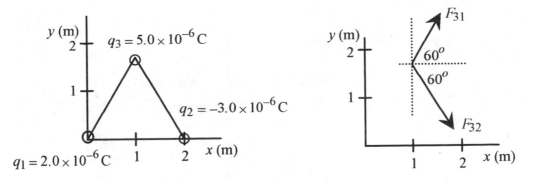

Solution: This is a more complicated vector problem. First calculate the force on q_3 due to q_1 and then due to q_2. Adding these vector forces is called **superposition**. The force on q_3 due to q_1 is

$$F_{31} = 9.0 \times 10^9 \frac{\text{Nm}^2}{\text{C}^2} \frac{(2.0 \times 10^{-6}\text{C})(5.0 \times 10^{-6}\text{C})}{4.0\,\text{m}^2} = 0.0225\,\text{N}$$

This force is directed as shown in the figure on the right above. Likewise, the force on q_3 due to q_2 is

$$F_{32} = 9.0 \times 10^9 \frac{\text{Nm}^2}{\text{C}^2} \frac{(3.0 \times 10^{-6}\text{C})(5.0 \times 10^{-6}\text{C})}{4.0\,\text{m}^2} = 0.0338\,\text{N}$$

This is directed as shown. Now the resultant force can be written. The x-component of these two forces is

$$R_x = F_{31}\cos 60^o + F_{32}\cos 60^o = (0.0112 + 0.0169)\text{N} = 0.0281\,\text{N}$$

The y-component of these two forces is

$$R_y = F_{31}\sin 60^o - F_{32}\sin 60^o = (0.0195 - 0.0292)\text{N} = -0.0097\text{N}$$

Now, to find the resultant, create a separate vector diagram and draw in the components.

$$R^2 = R_x^2 + R_y^2 = .0281^2 + (-.0097)^2 = .0297$$

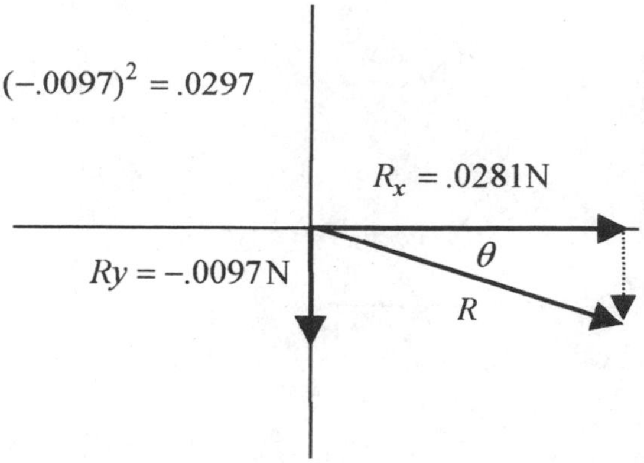

$R_x = .0281N$

$Ry = -.0097N$

θ

R

We found the magnitude of the resultant using the Pythagorean theorem. Now, just looking at the triangle in the diagram above, we can use any one of the trig functions to find the angle, θ.

$$\sin \theta = \frac{\text{opposite}}{\text{hypotenuse}} = \frac{.0097}{.0297} \Rightarrow \theta = \sin^{-1}\left(\frac{.0097}{.0297}\right) = 19.1^o$$

Remember that $\sin^{-1}$ is not 1/sin. It is the inverse sin function on your calculator. The angle θ could also be called -19°. Since angles below the positive x axis are usually negative, while angles counter clockwise from the +x axis are positive. By the way, we could have used cos or tan to get θ as well:

$$\cos\theta = \frac{\text{adjacent}}{\text{hypotenuse}} = \frac{.0281}{.0297} \quad \text{or} \quad \tan\theta = \frac{\text{opposite}}{\text{adjacent}} = \frac{.0097}{.0281}$$

Try the inverse operation with each of these functions with your calculator to make sure you are doing it correctly.

Look back at the magnitude and direction of the Resultant force. Does this make sense in the context of the problem? We had a positive charge positioned above and between another + charge on the left and a larger − charge on the right. So the resultant force should definitely be to the right and slightly down (the larger negative charge out pulls the upward push of the positive charge).

15-4 Calculate the electric field at the point P ($x = 3$, $y = 3$) due to two charges, $q_1 = 4.0 \times 10^{-6} C$ at the origin, and $q_2 = -3.0 \times 10^{-6} C$ at $x = 3.0$ m.

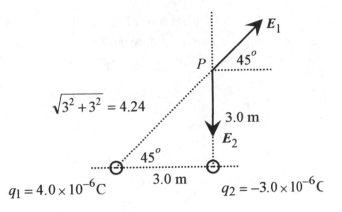

Solution: The electric field due to the first charge is

$$E_1 = k\frac{q}{r^2} = 9.0 \times 10^9 \frac{Nm^2}{C^2} \frac{4.0 \times 10^{-6}C}{18m^2} = 2.0 \times 10^3 \, N/C$$

Note the direction of E_1 as indicated in the diagram. The electric field due to the second charge is

$$E_2 = k\frac{q}{r^2} = 9.0 \times 10^9 \frac{Nm^2}{C^2} \frac{3.0 \times 10^{-6}C}{9.0m^2} = 3.0 \times 10^3 \, N/C$$

Now break the two electric field vectors up into components to find the components of the resultant.

$E_{1y} = 2.0 \times 10^3 \, N/C(\sin 45^0) = 1.41 \times 10^3 \, N/C$

$E_{1x} = 2.0 \times 10^3 \, N/C(\cos 45^0) = 1.41 \times 10^3 \, N/C$

$E_2 = 3.0 \times 10^3 \, N/C$

$$R_x = E_{1x} = 1.41 \times 10^3 \, N/C$$

$$R_y = E_{1y} - E_2 = 1.41 \times 10^3 \, N/C - 3.0 \times 10^3 \, N/C = -1.59 \times 10^3 \, N/C$$

The minus sign for E_2 is just to show **direction** of the field. Whenever you see a minus sign in front of an electric field or force, it is showing direction (unlike the minus sign in front of charge, which is telling you the type of charge). Now we are ready to find the Resultant electric field:

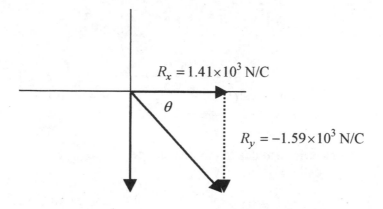

Now using your calculator, verify that the Pythagorean theorem gives 2.1×10^3 N/C for the Resultant electric field and that the angle θ in the diagram is 48° below the $+x$ axis.

15-5 A proton is placed in an electric field of 0.01 N/C (a) What is the force on the proton and (b) what is the magnitude of the proton's acceleration?

Solution: (a) The charge on a proton is the same as the charge on an electron (just positive instead of negative).

$$F = qE = 1.6 \times 10^{-19} \text{C} (0.01 \text{N/C}) = 1.6 \times 10^{-21} \text{N}$$

To find the acceleration, we use:

$$F = ma \Rightarrow 1.6 \times 10^{-21} \text{N} = 1.67 \times 10^{-27} \text{kg} (a) \Rightarrow a = 9.58 \times 10^5 \text{m/s}^2$$

The mass of a proton from the table in your text

15-6 Find an expression for the electric field due to a collection of charges of equal magnitude all placed in a line and separated by a distance d. The charges have alternate signs and the field to be calculated is the field on a line normal to the line of charges and opposite the central positive charge. Additionally, look for approximate expressions for long distances away from the line of charges.

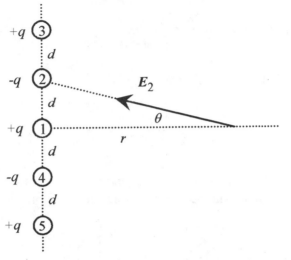

Solution: There are two things about this problem that you may have to start getting used to, depending upon your instructor. First, we are not plugging in numbers. This is the kind of problem that is solved just by handling the variables. If you are used to always dealing with numbers, this may take some time to get used to. Also, the key to solving these types of electric charge problems with variables only is to look for **symmetry** in the problem. If your instructor never does problems like this then don't get bogged down with this problem and move on to the next chapter. The first time you read this problem, don't worry about following every mathematical detail. Look at the big picture of how the symmetry of this problem is being treated.

The electric fields are labeled to correspond with the charges. The fields at a point r distance away from the line of charge are:

$$E_1 = k\frac{q}{r^2} \qquad\qquad E_2 = E_4 = k\frac{(-q)}{r^2 + d^2} \qquad\qquad E_3 = E_5 = k\frac{q}{r^2 + 4d^2}$$

Look at the geometry associated with E_2 and notice that this vector can be written in terms of a horizontal and a vertical component. The vector E_4 also has vertical and horizontal components. Do you see that the vertical components of E_2 and E_4 add to zero? The horizontal components add along r. The same is true of vectors E_3 and E_5. The horizontal component of E_2 is E_2 times the cosine of the angle θ, thus the contribution of E_2 and E_4 is two times the magnitude of E_2 times the cosine of the angle θ.

E_2 and E_4 contribute $2E_2 \cos\theta = k\dfrac{-2q}{r^2+d^2}\dfrac{r}{\sqrt{r^2+d^2}}$ pointed in along r.

Similarly E_3 and E_5 contribute $k\dfrac{2q}{r^2+d^2}\dfrac{r}{\sqrt{r^2+4d^2}}$ also pointed out along r

Combining the contributions from all the charges

$$E = kq\left[\frac{1}{r^2} - \frac{2r}{(r^2+d^2)^{3/2}} + \frac{2r}{(r^2+4d^2)^{3/2}}\right]$$

If r is comparable to d then this formula is appropriate. However if r is large compared to d then the r terms predominate in the denominators and the expression is approximately

$$E \approx kq\left[\frac{1}{r^2} - \frac{2}{r^2} + \frac{2}{r^2}\right] \approx k\frac{q}{r^2}$$

This formula shows that at long distances the array looks like one positive charge. That's because the array contains one extra positive charge and beyond that, an equal number of negative and positive charges.

16

ELECTRIC POTENTIAL
AND CAPACITANCE

Things to Remember

- In the first semester of physics we defined Work as force times displacement (or distance). So, for a force due to an electric field (qE), the formula for energy is $\Delta PE = -W = qEd$. See problem 16-1.

- Changes in voltage are expressed as $\Delta V = \dfrac{\Delta PE}{q}$. The unit of Volts = Joules/Coulomb.

- The definition of capacitance is $C = \dfrac{q}{V}$. Farads $= \dfrac{\text{Coulombs}}{\text{Volts}}$.

- The capacitance of a parallel plate capacitor (with the plates separated by air) is $C = \varepsilon_0 \dfrac{A}{d}$. The value of $\varepsilon_0 = 8.85 \times 10^{-12}\,\mathrm{C^2/(N \cdot m^2)}$. A capacitor with a dielectric is $C = \kappa \varepsilon_0 \dfrac{A}{d}$, where κ is the dielectric constant.

- The equivalent capacitance of capacitors in parallel is $C_{eq} = C_1 + C_2 + C_3 \ldots\ldots$ At the same time you memorize this, please also remember that the voltage on capacitors in parallel is the same, but the charge on each capacitor is different. This is true if the capacitors are attached to a single voltage source.

- The equivalent capacitance of capacitors in series is $\dfrac{1}{C_{eq}} = \dfrac{1}{C_1} + \dfrac{1}{C_2} + \dfrac{1}{C_3}\ldots\ldots$ Also memorize that for capacitors in series (attached to a single voltage source), the charge is the same, but not the voltage.

- The energy stored in a capacitor is $(1/2)CV^2$.

16-1 A charge of $+200\mu C$ is pushed a distance of 1.5 meters against an electric field of $13\,\text{N/C}$. (a) How much work is done on the charge? (b) What is the change in electric potential energy of the charge? (c) What is the change in voltage that the charge is pushed through?

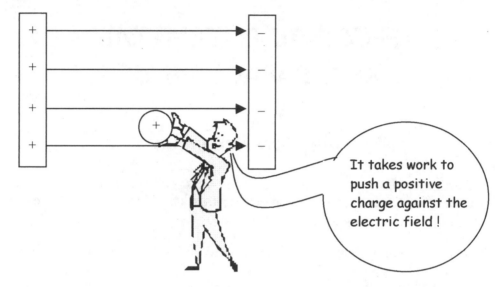

It takes work to push a positive charge against the electric field !

Solution: For starters, $+200\mu C$ means 200 microCoulombs or 200×10^{-6} Coulombs. Also, electric field lines run from the + charge to the − charge as shown above for the electric field between two parallel plates. Since like charges repel, a positive charge will naturally move in the same direction as the electric field lines. Pushing it against the electric field requires work, much like rolling a ball up a hill would require work against the gravitational field.

(a) Let's just use our knowledge (or use this example as a way of refreshing our knowledge) of the units of force and work to answer part (a). We have an electric field of $13\,\text{N/C} \times 200\times10^{-6}\,\text{C} = 2.6\times10^{-3}\,\text{N}$. This looks like the force we are going to need to push that charge against the field, doesn't it? We know that Work = Force × Distance, so $2.6\times10^{-3}\,\text{N}\times1.5\,\text{m} = 3.9\times10^{-3}\,\text{J}$. Remember that Newtons × meters = Joules.

(b) So $3.9\times10^{-3}\,\text{J}$ is the amount of work done on the charge and that work **goesinto** increasing the electric potential energy of that charge. The change in potential energy is positive. We put work in, and pushed the charge into a position where if we release the charge, it has farther to "fall" to the other negatively charged plate. It has more potential to gain kinetic energy when released for example, or it has more potential to do work. If you don't remember the **goesinto** concept for energy flow from earlier in this book, please review chapter 7 on work and energy. Don't worry about the numbers, just read a few of the problems to see the method of analyzing energy flow that we are talking about here.

Unfortunately, many textbooks like to play games with minus signs and write formulas that look like this: $\Delta PE = -W$. Don't get bogged down by this minus sign. Work done **on the charge** by an outside force **against the electric field** is negative. Work done **by the electric field** on the charge would be positive. The important thing for you to see in this problem is that the potential energy is increased. We did work on the charge and that went into increasing its potential energy. Be sure this makes physical sense to you before going on.

(c) The formula for voltage is

$$\Delta V = \frac{\Delta PE}{q} = \frac{3.9 \times 10^{-3} \text{J}}{200 \times 10^{-6} \text{C}} = 19.5 \text{ Volts}$$

This is also positive. High voltage means closer to the + side for a positive charge.

The units of electric field are N/C, but that is also equal to V/m. Therefore:

$$13 \text{ N/C} \implies 13 \text{ V/m} \times 1.5 \text{ m} = 19.5 \text{ V}$$

16-2 A charge of $q_1 = -6\mu C$ is located 2.0 cm above a charge of $q_2 = +3\mu C$ on the y-axis. (a) What is the potential half way between the charges at the point $y = 1$ cm? (b) At what two points along the y-axis is the potential equal to 0?

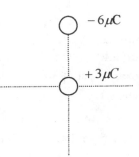

Solution: (a) The formula for the potential at a point in space due to a point charge is $V = k\dfrac{q}{r}$. The value of r is 1 cm or 0.01 m for both. We need to calculate the potential due to both charges and then add them together.

$$V_1 = \frac{9 \times 10^9 \text{ N} \cdot \text{m}^2}{\text{C}^2} \frac{-6 \times 10^{-6} \text{C}}{.01\text{m}} = \frac{-5.4 \times 10^6 \text{ N} \cdot \text{m}}{\text{C}} = \frac{-5.4 \times 10^6 \text{ J}}{\text{C}} = -5.4 \times 10^6 \text{ V}$$

$$V_2 = \frac{9 \times 10^9 \text{ N} \cdot \text{m}^2}{\text{C}^2} \frac{+3 \times 10^{-6} \text{C}}{0.01 \text{ m}} = \frac{+2.7 \times 10^6 \text{ N} \cdot \text{m}}{\text{C}} = +2.7 \times 10^6 \text{ V}$$

$$V_{at\ y=1cm} = -5.4 \times 10^6 \text{ V} + 2.7 \times 10^6 \text{ V} = -2.7 \times 10^6 \text{ V}$$

(b) Looking at the r in the denominator of the formula for potential, realize that the closer you are to a positive charge, the smaller the r in the denominator, and the larger the positive potential. The closer you are to a negative charge, the higher the negative potential. What we are looking for are two points where the positive and negative potentials will add to 0. One of these points should be between the charges, and the other point must be below the positive charge. Do you see why? The second point can't be above the negative charge because the negative charge is stronger than the positive charge. Any point above the negative charge would always be closer to the stronger charge, so there's no way of both potentials adding to 0.

$$k\frac{-6\mu C}{r_1} + k\frac{3\mu C}{r_2} = 0$$

Just to keep this straight, by r_1 we mean the absolute value of the distance from the negative charge q_1 to the point that we are looking for. By r_2 we mean the absolute value of the distance from the positive charge to the point we are looking for. Divide both sides of the equation by k ($0/k = 0$). The equation above reduces to

$$\frac{6}{r_1} = \frac{3}{r_2} \quad \Rightarrow \quad 3r_1 = 6r_2 \quad \Rightarrow \quad r_1 = 2r_2$$

Now comes the tricky part for those that are not mathematically inclined. By some miracle it needs to dawn on you that in order to find the point in between the charges, we need to set $r_1 + r_2 = 2$ cm and that will give us 2 equations and 2 unknowns. Do you see why? For example with just the equation above, we could plug in $r_1 = 1$cm and then $r_2 = 2$ cm, but that is not one point. That is 1 cm away from the negative charge and 2 cm away from the positive charge. That's 2 different points.

Now when you have two equations and two unknowns, plug one equation into the other:

$$r_1 + r_2 = 2\,\text{cm} \quad \Rightarrow \quad 2r_2 + r_2 = 2\,\text{cm} \quad \Rightarrow \quad 3r_2 = 2\,\text{cm} \quad \Rightarrow \quad r_2 = 0.67\,\text{cm}$$

This corresponds to the point $y = + \mathbf{0.67}$ **cm.**

That's the first point. Can you come up with the other constraint equation that relates r_1 and r_2 for the point below the positive charge? Go back to the diagram on the previous page and think about it for a moment and try to come up with this relation before reading on.......

The answer is $r_1 = r_2 + 2\,\text{cm}$. The equation $r_1 = 2r_2$ still holds, so:

$2r_2 = r_2 + 2\,\text{cm} \implies r_2 = 2\,\text{cm}$, which means the second point is 2 cm below the positive charge or $y = \textbf{-2.0 cm}$.

Remember, if a problem asks, "What is the **potential**?" then it is asking for an answer in **Volts**. If the question is, "What is the **potential energy**?" then the answer is in **Joules**.

Capacitance

16-3 A parallel plate capacitor has an area for the plates of $10\,\text{cm}^2$. The plates are separated by 4 mm. If nylon ($\kappa = 3.4$) is used as a dielectric in the capacitor, (a) calculate the capacitance. (b) If the dielectric strength (electric breakdown) of nylon is 14×10^6 V/m, how much charge can be placed on the capacitor?

Solution: (a) $d = \dfrac{4\,\text{mm}}{} \dfrac{1\,\text{m}}{1000\,\text{mm}} = 4 \times 10^{-3}\,\text{m}$ $A = \dfrac{10\,\text{cm}^2}{} \dfrac{(1\,\text{m})^2}{(100\,\text{cm})^2} = 1 \times 10^{-3}\,\text{m}^2$

$$C = \kappa \varepsilon_0 \frac{A}{d} = 3.4 \times 8.85 \times 10^{-12}\,\text{C}^2/(\text{N} \cdot \text{m}^2)\frac{1 \times 10^{-3}\,\text{m}^2}{4 \times 10^{-3}\,\text{m}} = 7.5 \times 10^{-12}\,\text{F}$$

(b) The maximum amount of voltage that we can put on this capacitor is

$$\frac{14 \times 10^6\,\text{V}}{\text{m}} \times \frac{4 \times 10^{-3}\,\text{m}}{} = 5.6 \times 10^4\,\text{V}$$

To find the charge we can use

$$C = \frac{q}{V} \implies 7.5 \times 10^{-12}\,\text{F} = \frac{q}{5.6 \times 10^4\,\text{V}} \implies q = 4.2 \times 10^{-7}\,\text{C}$$

16-4 A $6.0\,\mu\text{F}$ capacitor and an $8.0\,\mu\text{F}$ capacitor are connected in series. A potential of $200\,\text{V}$ is placed across them. Find the charge and potential on each.

Solution: The circuit is as shown. Since the capacitors **are in series the voltage on each capacitor is different,** but the charge on each is the same. You are best off just memorizing this point; it will save you time on the test. The equivalent capacitance is found from

$$\frac{1}{C} = \frac{1}{C_1} + \frac{1}{C_2} = \frac{1}{6.0\,\mu\text{F}} + \frac{1}{8.0\,\mu\text{F}} = \frac{4}{24\,\mu\text{F}} + \frac{3}{24\,\mu\text{F}} = \frac{7}{24\,\mu\text{F}}$$

so the equivalent capacitance is $(24/7)\mu\text{F}$ and the equivalent circuit is as shown.

The charge is calculated from $q = CV = (24/7)\mu\text{F} \times 200\,\text{V} = 686\,\mu\text{C}$.

The voltage on each capacitor is

$$V_1 = \frac{q}{C_1} = \frac{686\mu}{6.0\mu}\frac{\text{C}}{\text{F}} = 114\text{V} \qquad V_2 = \frac{q}{C_2} = \frac{686\mu}{8.0\mu}\frac{\text{C}}{\text{F}} = 86\text{V}$$

The sum of these voltages is the $200\,\text{V}$ applied to the combination.

NOTE: This general approach of reducing circuits to simpler equivalents by applying the rules for lumped circuit elements is at the heart of circuit analysis.

16-5 A $12\,\mu$F capacitor and a $10\,\mu$F capacitor are connected in parallel across a 100 V battery. Find the charge and potential on each capacitor.

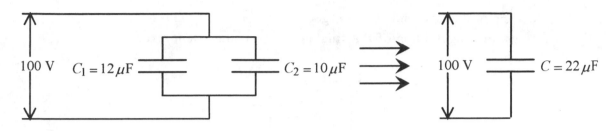

Solution: The voltage across each capacitor in parallel is the same, 100 V. The charge, however, depends on the capacitance. The fastest way to do capacitor combination problems on a test is to memorize these two points.

Speed

$$q_1 = C_1 V = 12\mu \quad \text{F}\times 100\text{V} = 12\times 10^{-4}\,\text{C} \quad q_2 = C_2 V = 10\mu \quad \text{F}\times 100\text{V} = 10\times 10^{-4}\,\text{C}$$

Look at the equivalence. Take this total charge and the applied voltage and find an equivalent capacitance.

$$C = \frac{q}{V} = \frac{22\times 10^{-4}\,\text{C}}{100\text{ V}} = 22\ \mu\text{F}$$

This is the sum of the capacitances as expected in a parallel arrangement.

16-6 The analysis of multiple capacitor combinations is done with (sometimes several) equivalent circuits. Consider the circuit shown that has parallel and series combinations.

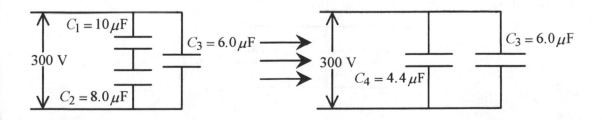

Solution: The first equivalent circuit comes from the combination of C_1 and C_2.

$$\frac{1}{C_4} = \frac{1}{C_1} + \frac{1}{C_2} = \frac{1}{10\,\mu\text{F}} + \frac{1}{8.0\,\mu\text{F}} = \frac{9.0}{40\,\mu\text{F}} \quad \text{or} \quad C_4 = 4.4\,\mu\text{F}$$

Now combine C_3 and C_4 to get $C_5 = C_3 + C_4 = 4.4\,\mu\text{F} + 6.0\,\mu\text{F} = 10.4\,\mu\text{F}$. This allows the construction of the final (simplest) equivalent circuit.

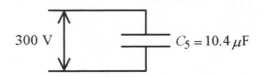

Now calculate the total charge $q_t = C_5 V = 10.4\,\mu\text{F} \cdot 300\,\text{V} = 3.1 \times 10^{-3}\,\text{C}$.

The charge on C_3 is $q_3 = C_3 V = 6.0\,\mu\text{F} \cdot 300\,\text{V} = 1.8 \times 10^{-3}\,\text{C}$.

The remaining charge is $1.3 \times 10^{-3}\,\text{C}$ and this is the charge on each of the other capacitors, C_1 and C_2. Remember that the charge on capacitors in series is the same. The voltage on each of these capacitors is

$$V_1 = \frac{q}{C_1} = \frac{1.3 \times 10^{-3}\,\text{C}}{10\,\mu\text{F}} = 130\,\text{V} \qquad V_2 = \frac{q}{C_2} = \frac{1.3 \times 10^{-3}\,\text{C}}{8.0\,\mu\text{F}} = 163\,\text{V}$$

The error here of these voltages not adding up to 300 V is about 2%, which is expected from round-off.

16-7 A $12\,\mu\text{F}$ capacitor is charged with a $200\,\text{V}$ source, then placed in parallel with an uncharged $7.0\,\mu\text{F}$ capacitor. Calculate the new voltage on each capacitor.

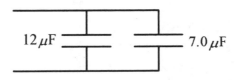

Solution: The $12\,\mu\text{F}$ capacitor initially placed on the $200\,\text{V}$ source has a charge of

$$q = CV = 12\,\mu\text{F} \cdot 200\,\text{V} = 2.4 \times 10^{-3}\,\text{C}$$

When connected to the $7.0\,\mu F$ capacitor the charge distributes so that the total charge remains the same and the new voltages are equal as they must be for capacitors in parallel. Take Q as the amount of charge that moves from the $12\,\mu F$ capacitor to the $7.0\,\mu F$ capacitor. The voltages as expressed by the ratio of q to C must be equal, so write

$$\frac{2.4\times10^{-3}\,C-Q}{12\,\mu F}=\frac{Q}{7.0\,\mu F} \quad \text{so} \quad Q=8.84\times10^{-4}\,C$$

The voltage on each of the capacitors is

$$V_7=\frac{q}{C}=\frac{8.84\times10^{-4}\,C}{7.0\mu \quad F}=126V \qquad V_{12}=\frac{q}{C}=\frac{1.52\times10^{-3}\,C}{12\mu \quad F}=126V$$

16-8 Go back to the previous problem of the $12\,\mu F$ capacitor charged to $200\,V$ and then connected to the $7.0\,\mu F$ capacitor and calculate (a) the amount of energy in the single capacitor before the second is placed in parallel, and (b) the energy stored in both capacitors after the charge has equalized across both.

Solution: (a) The $12\,\mu F$ capacitor at $200\,V$ has a stored energy of

$$U=(1/2)CV^2=(1/2)12\mu F(40{,}000\ V)^2=0.24J$$

NOTE: The units of energy are $\left(FV^2\right)=\left(\left(C/V\right)\left(J/C\right)^2\right)=\left(J\right)$.

(b) After the capacitors are connected, the voltage drops to $126\,V$ and the energy stored in the two capacitors is

$$U=\frac{1}{2}12\mu F(126\ V)^2+\frac{1}{2}7.0\mu \quad F(126\ V)^2=9.5\mu F(126\ V)^2=0.15\ J$$

Where did the energy go? The difference between the 0.24 J and the 0.15 J is the kinetic energy needed to transport the charge or the work to move the charge.

Insight

17

DIRECT CURRENT CIRCUITS

Current and Resistivity

17-1 A current of $5.0\,\text{A}$ exists in a wire of $0.50\,\text{cm}$ radius. (a) Calculate the current density. (b) Calculate the amount of charge that passes a cross section each second. (c) Calculate the number of electrons per second.

Solution:
$$J = \frac{I}{Area} = \frac{5.0\,\text{A}}{\pi(0.25 \times 10^{-4}\,\text{m}^2)} = 6.4 \times 10^4\,\frac{\text{A}}{\text{m}^2}$$

$$I = \frac{\Delta q}{\Delta t} \ \Rightarrow \ 5.0\,\text{A} = 5.0\,\text{C/s} \ \text{ so } 5.0\,\text{C} \text{ passes a cross section each second}$$

$$\frac{5.0\,\text{C}}{\text{s}} \ \frac{1\text{e}^-}{1.6 \times 10^{-19}\,\text{C}} = 3.1 \times 10^{19}\,\text{e}^- \text{ per second}$$

The **resistance** of a conductor is defined as the voltage divided by the current, $R = V/I$, and has a special unit, the ohm = volt/amp or in symbol form, $\Omega = V/A$. You will also see this equation written as $V = IR$. The **resistivity** of a material is usually given in a problem and is an inherent property of the material. The difference between the two is illustrated in the following problem.

17-2 The resistivity of copper is $1.7 \times 10^{-8}\,\Omega \cdot \text{m}$. What current flows through a $2.0\,\text{m}$ long copper conductor of $1.0 \times 10^{-4}\,\text{m}^2$ cross section when $20\,\text{V}$ is applied?

Solution: First calculate the resistance.

$$R = \rho \frac{length}{Area} = \frac{1.7 \times 10^{-8}\,\Omega \cdot \text{m}(2.0\,\text{m})}{1.0 \times 10^{-4}\,\text{m}^2} = 3.4 \times 10^{-4}\,\Omega$$

112

Now the current is $\qquad I = \dfrac{V}{R} = \dfrac{20\ \text{V}}{3.4 \times 10^{-4}\,\Omega} = 5.9 \times 10^4\ \text{A}$

Resistors in DC Circuits

In this section we will look first at series and parallel resistors, then at how series and parallel combinations are analyzed by writing successively simpler equivalent circuits.

Series Resistors

Take three resistors in series as shown in the accompanying diagram.

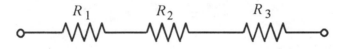

If a voltage, V, is applied to this combination the current through each resistor is the same. The voltage statement for each of the resistors is $V_1 = IR_1, V_2 = IR_2, V_3 = IR_3$.   The individual   voltages   across   each   resistor   must   add   to   the   total   so $V = V_1 + V_2 + V_3 = I(R_1 + R_2 + R_3)$.   The equivalent resistor for this combination is $R = R_1 + R_2 + R_3$. **Resistors in series add linearly**.

★
Remember

17-3 What is the equivalent resistance of $3.0\,\Omega$ and $4.0\,\Omega$ resistors placed in series? What is the current in each resistor if $10\,\text{V}$ is applied to the combination?

Solution: The equivalent resistance is $R_{eq} = 3.0\Omega + 4.0\Omega = 7.0\Omega$. The **current is the** **same in each resistor** and has value $I = V/R_{eq} = 10\text{V}/7.0\Omega = 1.4\text{A}$. Remember that if there is only one path in a circuit, then there is only one current.

Remember

Parallel Resistors

Place three resistors in parallel as shown below.   If a voltage is applied to this combination the voltage across each resistor is the same and the currents are determined by $V = I_1R_1$, $V = I_2R_2$, and $V = I_3R_3$.   These currents must add up to the total current delivered to the combination.

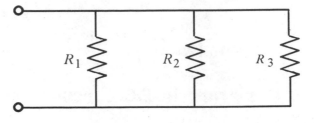

$$I = I_1 + I_2 + I_3 = \frac{V}{R_1} + \frac{V}{R_2} + \frac{V}{R_3} = V\left(\frac{1}{R_1} + \frac{1}{R_2} + \frac{1}{R_3}\right) = \frac{V}{R_{eq}}$$

An equivalent resistor (for this combination) then would have the form

Remember

$$\frac{1}{R_{eq}} = \frac{1}{R_1} + \frac{1}{R_2} + \frac{1}{R_3}$$

Resistors in parallel add reciprocally.

17-4 What is the current in each resistor for a parallel combination of $3.0\,\Omega$, $3.5\,\Omega$, and $4.0\,\Omega$ resistors with 8.0 V?

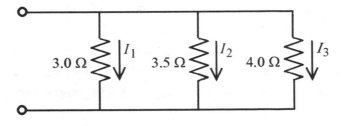

Solution: $I_1 = \dfrac{8.0\text{ V}}{3.0\,\Omega} = 2.7\text{ A}$ $I_2 = \dfrac{8.0\text{ V}}{3.5\,\Omega} = 2.3\text{ A}$ $I_3 = \dfrac{8.0\text{ V}}{4.0\,\Omega} = 2.0\text{ A}$

The total current delivered by the power source is $I = 2.7\text{A} + 2.3\text{A} + 2.0\text{A} = 7.0\text{A}$. Now check the total by using the equivalence

$$\frac{1}{R_{eq}} = \frac{1}{3.0\,\Omega} + \frac{1}{3.5\,\Omega} + \frac{1}{4.0\,\Omega} = \frac{36.5}{42\,\Omega} \text{ or } R_{eq} = 1.15\,\Omega$$

Using the equivalent resistance ($1.15\ \Omega$) the total current is

$$I = V/R_{eq} = 8.0V/1.15\Omega = 7.0A$$

Remember that if there is more than one path in the circuit, there is more than one current. Further, more of the current will go through the path of lesser resistance.

Insight

Series-Parallel Combinations

17-5 Calculate the equivalent resistance of the series-parallel combination shown below.

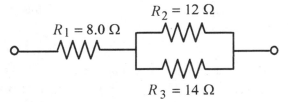

$R_2 = 12\ \Omega$

$R_1 = 8.0\ \Omega$

$R_3 = 14\ \Omega$

Solution: The first step in this circuit analysis is to find the equivalent of the parallel combination.

$$\frac{1}{R_4} = \frac{1}{R_2} + \frac{1}{R_3} = \frac{1}{12\Omega} + \frac{1}{14\Omega} = \frac{13}{84\Omega} \quad \text{or} \quad R_4 = \frac{84\Omega}{13} = 6.5\Omega$$

The circuit then becomes two series resistors and they add to a single resistor.

NOTE: The equivalent resistance of any two resistors in parallel is lower than either of the two alone.

Insight

Now adding the two resistors in series, we get

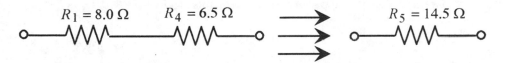

$R_1 = 8.0\ \Omega$ $R_4 = 6.5\ \Omega$ $R_5 = 14.5\ \Omega$

17-6 Apply $20\,V$ across the network of resistors shown originally in the previuos problem and find the voltage across each resistor and current through each resistor.

Solution: In order to find the current in a resistor in the original network, it is necessary to work back through the equivalent circuits. Start with the last circuit. Apply 20 V to a 14.5 Ω resistor for a current of $I = V/R = 20\,V/14.5\Omega = 1.4\,A$. Working backward, this current through R_1 produces a voltage across this resistor of

$V_1 = (1.4\,\text{A})R_1 = (1.4\,\text{A})(8.0\,\Omega) = 11\,\text{V}$. This current passes through the equivalent resistor R_4 producing a voltage across it of $V_4 = (1.4\,\text{A})R_4 = (1.4\,\text{A})(6.5\,\Omega) = 9.1\,\text{V}$. V_1 and V_4 add to 20 V, the total applied, to within round-off error.

The 1.4 A divides between R_2 and R_3. The voltage across R_2 and R_3 is $V_4 = 9.1\,\text{V}$ so

$$I_2 = 9.1\,\text{V}/12\,\Omega = 0.76\,\text{A} \quad I_3 = 9.1\,\text{V}/14\,\Omega = 0.65\,\text{A}$$

Again this total adds to 1.4 A, within round-off error. This analysis produces the voltage across and current through each resistor in the network. The numbers are internally consistent (the currents through R_2 and R_3 add up to the current through their equivalence) giving confidence in the calculation.

17-7 In the circuit below, calculate the voltage across, current through, and power requirements of the $8.0\,\Omega$ resistor when $15\,\text{V}$ is applied to the network.

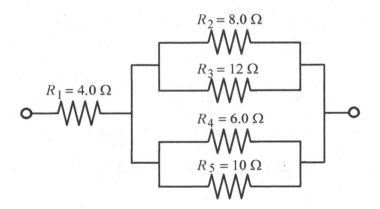

Solution: Go after the ones in parallel first. The equivalent of R_2 and R_3 is

$$\frac{1}{R_6} = \frac{1}{R_2} + \frac{1}{R_3} = \frac{1}{8.0\,\Omega} + \frac{1}{12\,\Omega} = \frac{5}{24\,\Omega} \quad \text{or} \quad R_6 = \frac{24}{5} = 4.8\,\Omega$$

A similar analysis on R_4 and R_5 yields

$$\frac{1}{R_7} = \frac{1}{R_4} + \frac{1}{R_5} = \frac{1}{6.0\,\Omega} + \frac{1}{10\,\Omega} = \frac{8}{30\,\Omega} \quad \text{or} \quad R_6 = \frac{30}{8} = 3.8\,\Omega$$

Now the equivalent can be drawn.

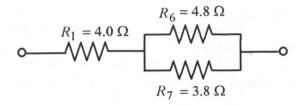

The parallel combination of R_6 and R_7 produces another equivalent.

$$\frac{1}{R_8} = \frac{1}{R_6} + \frac{1}{R_7} = \frac{1}{4.8\Omega} + \frac{1}{3.8\Omega} = \frac{8.5}{18\Omega} \text{ or } R_8 = \frac{18}{8.5} = 2.1\Omega$$

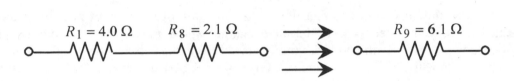

The total current is $I = V/R_9 = 15\,\text{V}/6.1\Omega = 2.46\,\text{A}$.

This current through the 2.1Ω resistor produces

$$V_8 = (2.46\,\text{A})R_8 = (2.46\,\text{A})(2.1\Omega) = 5.16\,\text{V}$$

Look back through the equivalent circuits and notice that this voltage is across $R_8, R_6,$ and R_2. The current through R_2 then is

$$I_2 = V_8/R_2 = V_2/R_2 = 5.16\,\text{V}/8.0\,\Omega = 0.64\,\text{A}.$$

The power dissipated in this resistor is $I^2R = (0.64\text{A})^2(8.0\Omega) = 3.3\,\text{W}$.

The general procedure for circuit analysis is to look at the circuit and find parts of it that can be replaced with simpler equivalents and just keep applying the process until the circuit is reduced to one resistor, perform $V = IR$ on that resistor, and move back through the equivalents to obtain the desired information. **Pattern** ·

Kirchhoff's Laws

Kirchhoff's laws, applied to circuits containing multiple branches, voltage sources, and resistors allow calculation of currents in each branch of the circuit. Start with a very simple circuit consisting of a battery and resistor. The battery polarity is taken as indicated below with a positive charge leaving the positive plate. The current is as indicated by the arrow. An amount of charge or current over time starting at the negative plate of the battery is viewed as gaining energy in passing through the battery and loosing this same amount of energy in the resistor.

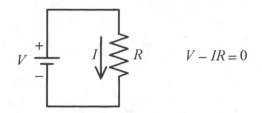

$$V - IR = 0$$

By the way, the direction of current flow is the direction of the flow of positive charge. Actually, it is the electrons, or negative charge which move. Back before people understood electricity, somebody arbitrarily decided to make current flow the direction of the flow of positive charge. Since then, everybody has been too lazy or too stubborn to break tradition and change.

Kirchhoff Voltage Loop Equations

The $V - IR$ equation for this simple circuit illustrates the **First Kirchhoff Law**: The algebraic sum of the changes in potential (voltage) encountered in a complete traverse of a path is zero. This is analogous to a conservation of energy statement.

Remember

The only difficulty in applying Kirchhoff's laws is keeping the algebraic signs correct. When confused, go back to this simple diagram of a battery and resistor as shown above and the current flowing (in the external circuit) from the positive side of the battery through the resistor to the negative side of the battery and write the $V - IR = 0$ statement establishing that **when a battery is traversed in a positive direction there is a gain in energy giving V a plus sign, and when a resistor is traversed in the direction of the current arrow there is a loss in energy giving IR a negative sign.** Every Kirchhoff law problem you do should have this little circuit drawn in a corner of the paper to remind you of the algebraic signs and their importance.

Insight

Think of a "piece of charge" going through the circuit and gaining energy as it passes through the battery and loosing energy as it passes through the resistor. In this simple circuit, the direction of the current is clear. In multiple-branch circuits, it is not possible to "guess" the correct directions for the currents. Make the best educated guess that you can and let the mathematics tell you whether the currents are positive or negative.

17-8 Consider the three-branch circuit below. Find the current through each resistor.

Solution: Make your best estimate of the current direction for each resistor and draw and label the current arrows. It is not necessary to have the direction correct, it is only necessary to be consistent in applying the sign convention. Now make two loops around this circuit, one around the left and one around the right as indicated by the closed loops. Remember, when traversing a battery from minus to positive write $+V$, and when traversing the resistor in the direction of the current arrow write $-IR$.

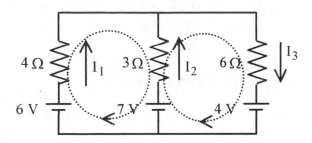

Around the first loop, we calculate $\quad 6 - 4I_1 + 3I_2 - 7 = 0$

In traversing this loop start at the low, or negative, side of the 6 V battery and proceed around the loop as follows:

- 6 is positive
- $4I_1$ is negative because the resistor is traversed in the direction of the current arrow
- $3I_2$ is positive because the resistor is traversed in the direction opposite to the current arrow
- 7 is negative because the battery is traversed from plus to minus

Around the second loop, we calculate

$$7 - 3I_2 - 6I_3 - 4 = 0$$

Rewrite these two equations as

$$-4I_1 + 3I_2 \qquad = 1$$
$$3I_2 + 6I_3 = 3$$

Writing the equations this way illustrates very clearly that they cannot be solved. In mathematical language they are two equations in three unknowns. The third equation necessary to solve for the three currents comes from the next Kirchhoff law.

Kirchhoff Current Junction Equations

The **Second Kirchhoff Law** deals with the currents: **the sum of the currents to any junction must add to zero**. This law is often forgotten. It is, however, easily remembered via a whimsical law called the "fat wire law." The fat wire law is very Remem simple. The sum of the currents to any junction must equal zero, otherwise the wire will get fat. Applying this second Kirchhoff law to the junction just above the 3Ω resistor, we get

$$I_1 + I_2 - I_3 = 0$$

Now we have three <u>independent</u> equations in three unknowns.

NOTE: For convenience in the discussion, individual equations in a set of equations will be designated by letter symbols. The number of significant figures used in the calculations will be kept to a minimum so as not to obscure the mathematical procedure with numbers. Likewise, units will not be used. Two techniques for solving multiple equations in multiple unknowns will be presented. It is not necessary to know both these techniques. Pick a method that you feel comfortable with and learn it well.

$$a : I_1 + I_2 - \quad I_3 = 0$$
$$b : -4I_1 + 3I_2 \quad = 1$$
$$c : \quad 3I_2 + 6I_3 = 3$$

At this point you may be asking, "Why didn't we take another loop to obtain a third equation?" Taking another loop in the circuit will only serve to generate another equation which is a linear combination of the first two loop equations. Take a loop around the outside of the circuit and write

$$6 - 4I_1 - 6I_3 - 4 = 0 \quad \text{or} \quad 4I_1 + 6I_3 = 2$$

This equation is just the difference of the first two loop equations and is not a linearly independent third equation. Therefore it cannot be used to solve for the three currents. As an exercise, take a figure-eight path around the circuit and prove that this statement is also a linear combination of the first two loop equations.

Now to the solution of these three equations. For those of you who need some algebra review, let's go over two general approaches to solving multiple equations with multiple unknowns.

1. The most popular and direct approach is to add and subtract equations. The idea is to produce two equations in two unknowns and then further add and subtract to solve for one current and work backwards to find all the currents.

multiply $4 \times$ a: $\quad 4I_1 + 4I_2 - 4I_3 = 0$

add b: $\quad -4I_1 + 3I_2 \qquad = 1$

to obtain $\quad \alpha$: $\qquad 7I_2 - 4I_3 = 1$

Combine equation α with 2/3 of equation c and write α and β.

$$\alpha: \quad 7I_2 - 4I_3 = 1$$
$$\beta: \quad 2I_2 + 4I_3 = 2$$
$$9I_2 = 3 \quad \text{or} \quad I_2 = 1/3$$

Place $I_2 = 1/3$ into c. $\qquad 3(1/3) + 6I_3 = 3$ or $I_3 = 1/3$

Now place $I_2 = 1/3$ and $I_3 = 1/3$ into a. $\qquad I_1 + (1/3) - (1/3) = 0$ or $I_1 = 0$

Check these current values in each of the original equations to verify that they are correct.

2. Here's another approach to solving the original set of equations:

$$I_1 + I_2 - I_3 = 0$$
$$-4I_1 + 3I_2 \quad = 1$$
$$I_2 + 2I_3 = 1$$

Multiply the first equation by 4 and add to the second equation generating a new second equation.

$$I_1 + I_2 \quad - I_3 = 0$$
$$7I_2 - 4I_3 = 1$$
$$I_2 + 2I_3 = 1$$

This makes the coefficient of I_1 equal to zero in the second and third equations, producing two equations in two unknowns. Now multiply the second equation by $-1/7$ and add to the third equation generating a new third equation.

$$I_1 + I_2 - I_3 = 0$$
$$7I_2 - 4I_3 = 1$$
$$(18/7)I_3 = 6/7$$

Now solve directly with $I_3 = 1/3$. Substituting this into the second equation, $I_2 = 1/3$. Putting these two values into the first equation obtains $I_1 = 0$.

Now go back to the equations where $18I_3 = 6$ or $I_3 = 1/3$.
Put $I_3 = 1/3$ into $7I_2 - 4I_3 = 1$ to obtain $I_2 = 1/3$.
Put $I_2 = 1/3$ and $I_3 = 1/3$ to obtain $I_1 = 0$.

Both of the above techniques work. Pick the one that fits you and the set of equations you are solving best and proceed carefully. Also, on a test, don't consume all your time with algebra. If you get bogged down, make sure you go on to other problems on the test. Practive solving these types of problems shortly before the test in order to build up your speed. (By "shortly before the test," we mean days, not minutes.)

Pattern

The procedure for solving Kirchhoff's laws problems is the following:

1. Draw the battery and resistor circuit in the corner of your paper to remind you of the sign convention.

2. Place current arrows next to each resistor and label them.

3. Pick a junction and write the current statement.

4. Draw loops and write the loop equations.

5. Solve for the currents and check them in the original equations.

17-9 Solve for the current in each of the resistors in the circuit shown below.

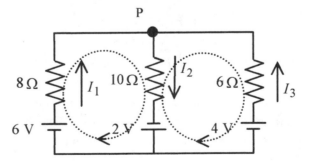

Solution: Follow the procedure for Kirchhoff problems.   This is a subject that is conceptually easy but because of the large number of manipulations it is often hard to actually do.   Work this problem as you are following along in the book.   Draw the circuit and begin.

•   Draw the single battery and single resistor with current arrow and write the $V - IR = 0$ equation.

- Place and label the current arrows.

- Write the statement of Kirchhoff's second law for the top center of the circuit labelled point P.

$$I_1 + I_3 = I_2$$

- Write a loop statement for the left side of the circuit.

$$6 - 8I_1 - 10I_2 + 2 = 0$$

- Write a loop statement for the right side of the circuit.

$$-2 + 10I_2 + 6I_3 - 4 = 0$$

- Now write these three equations in a convenient form for solution.

$$I_1 \quad - I_2 \ + I_3 = 0$$
$$8I_1 + 10I_2 \qquad = 8$$
$$10I_2 + 6I_3 = 6$$

As an exercise solve these equations and verify that the solutions are $I_1 = 17/47 = 0.36$, $I_2 = 24/47 = 0.51$, and $I_3 = 7/47 = 0.15$ respectively. Check these answers in the original equations.

17-10 Consider a more complicated problem as shown below and calculate the currents in the resistors.

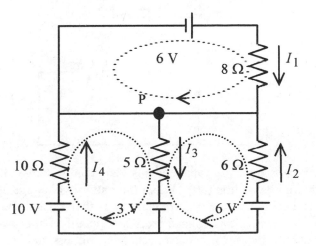

Solution: Draw the battery and resistor circuit, and $V - IR$ equation to remind you of the algebraic sign convention.

Note the current arrows. Remember that if one of the currents comes out negative it means only that the current is opposite to the direction of your arrow.

Apply Kirchhoff's second law (adding currents to 0 at a juntion) to the point labelled P at the top of the three resistors.

$$I_1 + I_2 + I_4 = I_3$$

The loop around the top branch of the circuit is $6 - 8I_1 = 0$.

A loop around the lower left branch is $10 - 10I_4 - 5I_3 - 3 = 0$.

A loop around the lower right branch is $3 + 5I_3 + 6I_2 - 6 = 0$.

Rewrite the equations in convenient form for solving.

$$
\begin{aligned}
I_1 + I_2 - I_3 + I_4 &= 0 \\
8I_1 &= 6 \\
6I_2 + 5I_3 &= 3 \\
5I_3 + 10I_4 &= 7
\end{aligned}
$$

Immediately $I_1 = 3/4 = 0.75$. Practice your algebra and verify that the other currents are $I_2 = -0.20$, $I_3 = 0.84$, and $I_4 = 0.28$.

RC Circuits

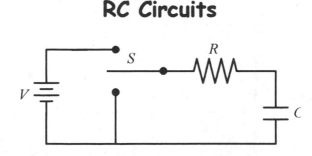

For the circuit above, assume the capacitor has zero charge and voltage, and place the switch (S) in the charging position (up). When the battery voltage is applied to the R and C in series, current begins to flow. Current through the resistor causes a voltage drop across the resistor. Because of this voltage drop, less than the battery voltage is applied to the capacitor. As current flows, the capacitor charges and less and less current flows until there is no current, no voltage across the resistor, and the capacitor is charged to the battery voltage ($Q = CV$).

The charge on the capacitor increases with time starting with zero charge and eventually reaching maximum charge of CV following an exponential function.

$$q = CV\left(1 - e^{-t/RC}\right)$$

This function fits our understanding of how the circuit operates since when $t = 0$, q is zero. This is because $e^0 = 1$. (Remember that any number raised to the 0th power is equal to 1.) Also, when t is very large, $q = CV$, the capacitor is totally charged.

The voltage on the capacitor is $q/C = V_C = V\left(1 - e^{-t/RC}\right)$

Here the notation V_C denotes the time varying voltage on the capacitor.

The current in the circuit declines exponentially according to $i = (V/R)e^{-t/RC}$

In biological systems that grow exponentially the systems are often characterized by giving the doubling time, the time for the system to double in number, size, or mass. In electrical systems that grow exponentially the systems are characterized by a time constant, the time to make the exponent of e equal to 1. The time constant for this circuit is RC. The units of RC are seconds. Note that there is a separate set of equations for discharging a capacitor. Here, we will do a set of problems for the charging case.

Insight

17-11 A $10\,k\Omega$ resistor and a $20\,\mu F$ capacitor are placed in series with a $12\,V$ battery. Refer to the circuit diagram at the beginning of this section. Find the charge on the capacitor, the current, and the voltages on the capacitor and resistor at the instant the switch is closed, $t = 0$.

Solution: At $t = 0$, the charge on the capacitor is zero.

At $t = 0$, the current is $i = V/R = 12\,V/10 \times 10^3\,\Omega = 1.2 \times 10^{-3}\,A$.

At $t = 0$, the voltage on the capacitor is zero (it has no charge) and the entire battery voltage of $12\,V$ is across the resistor.

17-12 For the circuit of the previous problem, find the time constant, and the charge, current, V_R, and V_C at a time equal to one time constant.

Solution: The time constant is R $C = (10 \times 10^3\,\Omega) \cdot (20 \times 10^{-6}\,F) = 0.20s$.

The charge on the capacitor at $t = 0.20$ s is

$$q\big|_{t=RC} = CV\left(1 - e^{-1}\right) = 12\,\mathrm{V}(20 \times 10^{-6}\,\mathrm{F})0.63 = 1.5 \times 10^{-4}\,\mathrm{C}$$

The current at $t = 0.20$ s is

$$i\big|_{t=RC} = \frac{V}{R}e^{-1} = \frac{12\,\mathrm{V}}{1.0 \times 10^{4}\,\Omega}0.37 = 4.4 \times 10^{-4}\,\mathrm{A}$$

The voltage across the capacitor is $V_C = V\left(1 - e^{-1}\right) = 12\mathrm{V} \times 0.63 = 7.6\mathrm{V}$.

The voltage across the resistor is $12\mathrm{V} - 7.6\mathrm{V} = 4.4\mathrm{V}$.

These problems can be deceptively easy. Be sure you know how to manipulate the exponents on your calculator. Don't get a test problem wrong because you did not practice all the steps in the problem and were unfamiliar with manipulating exponents on your calculator.

Watch Out!

17-13 For the same circuit, how long does it take for the capacitor to reach 80% of its final charge?

Solution: To find the time for 80% charge, set q equal to 80% of the final charge, or $q = 0.80\,CV$, and solve for t.

$$0.80CV = CV\left(1 - e^{-t/RC}\right) \;\Rightarrow\; 0.80 = 1 - e^{-t/RC} \;\Rightarrow\; e^{-t/RC} = 0.20$$

For convenience, switch to positive exponents so

$$1/e^{t/RC} = 0.20 \;\Rightarrow\; e^{t/RC} = 1/0.20 = 5$$

In order to solve for t, switch the exponential equation to a logarithmic equation. One of the functions of logarithms is to solve for variables in exponents. Algebraically, we take the natural log of both sides of the equation (and we know that $\ln(e^x) = x$).

$$t/RC = \ln 5 \quad \text{or} \quad t = RC\ln 5$$

Now put in the values for R and C.

$$t = RC\,\ln 5 = (1.0 \times 10^{4}\,\Omega) \cdot (20 \times 10^{-6}\,\mathrm{F}) \cdot \ln 5 = 0.32\mathrm{s}\,.$$

As a check, note that $1 - e^{-t/RC} = 1 - e^{-0.32/0.20} = 0.80$.

18

MAGNETISM

Things to Remember

There are lots of different formulas for different situations in magnetism. Write down the formula for each specific case on your equation sheet.

- The magnetic force on a point charge moving in a magnetic (B) field is
  $F = qvB\sin\theta$. Force is always in Newtons which is $\text{kg}\cdot\text{m/s}^2$. The magnetic field is in Tesla and velocity in meters/second.
- The force on a current-carrying wire is $F = BIL\sin\theta$. Length is in meters.
- A current carrying loop of wire in a magnetic field experiences a torque, given by
  $\tau = BIA\sin\theta$.
- The magnetic field produced by a current-carrying wire is given by Ampere's Law:

  $B = \dfrac{\mu_0 I}{2\pi r}$, with $\mu_0 = 4\pi \times 10^{-7}\,\text{T}\cdot\text{m/A}$. Just take a moment to look at this equation

  and the units of μ_0 . Do you see how the units of μ_0 cancel with all the other units on the right side of the equation, so that the final units are Tesla for magnetic field?
- The magnetic force between two parallel conductors is $F = \dfrac{\mu_0 I_1 I_2}{2\pi d}$.

- The magnetic field produced by a circular loop of wire carrying a current is $B = \dfrac{\mu_0 I}{2R}$.

- The magnetic field inside a solenoid is $B = \mu_0 n L$, where n is the number of turns per unit length.

18-1 A proton is moving perpendicular to a magnetic field of 3×10^4 Gauss directed toward the top of this page. If the force on the proton is 2×10^{-11} N and directed into the page, what is the (a) direction and (b) magnitude of its velocity?

Solution: (a) Let's get the direction of the movement of the proton first. For this we use the right-hand rule. The diagram at the top of page 128 shows the right hand tilted slightly, so we can see all the directions. For practice, place your hand flat down next to this page, palm facing down. Your fingers should be pointed toward the top of the page. Fingers represent the direction of the magnetic field. The direction that your palm faces is the direction of the force. This is easy to remember, because if you were to push with your hand, the force would be in the direction of your palm. Your thumb (at a right angle to your fingers) points in the direction of the velocity of a **positive charge**. So the answer to the question is that the velocity of the proton is to the left side of the page.

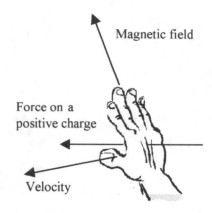

Magnetic field

Force on a positive charge

Velocity

The force that is on the proton is going to change the direction of its motion, so it will start to have a component of velocity into the page. To get an idea of directions as time passes, pick your hand up off the page and rotate it so that the fingers always point in the same direction, but the thumb rotates toward pointing into the page. Now the palm starts to indicate a direction of force more toward the right of the page. The charge would actually be moving during this time, so the result is that charges in a magnetic field **spiral**. We hope you didn't expect the charge to move in the same direction as the magnetic field – that's what happens in an electric field.

(b) To get the magnitude of the velocity, first convert Gauss to Tesla:

$$3 \times 10^4 \, \text{Gauss} \frac{10^{-4} \, \text{Tesla}}{1 \, \text{Gauss}} = 3 \, \text{T}$$

$$F = qvB \sin \theta = 2 \times 10^{-11} \, \text{N} = (1.6 \times 10^{-19} \, \text{C})v(3\text{T})(1) \implies v = 4.17 \times 10^7 \, \text{m/s}$$

Remember ★ θ is the angle between the magnetic field and the direction of the velocity. In this case $\theta = 90^o$ and $\sin 90^o = 1$. (Memorize the relation that $\sin 90^o = 1$.)

18-2 An electron is moving to the right side of the page, with a velocity of $2 \times 10^6 \, \text{m/s}$, directed 30° with respect to the magnetic field of 4 T as shown. Find (a) the direction and (b) the magnitude of the force on the electron.

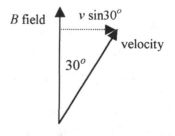

B field

$v \sin30^o$

velocity

30°

Solution: (a) In order to use the right-hand rule, realize that only the component of velocity perpendicular to the magnetic field (B field) is what will cause a force on the charge. This perpendicular component of the electron's velocity

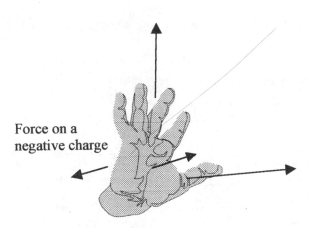

Force on a
negative charge

is directed to the right of the page, and would correspond to the velocity $\times \sin 30^o$. Place your right hand flat next to the page with fingers pointed to the top of the page along with the magnetic field and thumb pointed right. In this case your palm faces up. That would be the force on a positive charge. The direction of a force on a negative charge is in the opposite direction, or into the page. This is the best way to handle negative charges: Use the right hand rule to find the direction of the force on a positive charge and then you know that the force on the negative charge is in the opposite direction.

(b) For the magnitude of the force, we calculate

$$F = qvB\sin\theta = F = (1.6\times10^{-19}\,\text{C})(2\times10^6\,\text{m/s})(4\text{T})(\sin 30^o) = 6.4\times10^{-13}\,\text{N}$$

18-3 Find the force on an electron (a) at rest in a magnetic field of 2T, and (b) moving with a velocity of 1 million, billion, trillion meters per second in the same direction as the magnetic field.

Solution: The answer to both parts is 0. If the charge has no velocity it does not experience any force due to the magnetic field. Also, if it moves parallel with the field, it experiences no force.

18-4 A wire with a current of 15 A is positioned perpendicular to a magnetic field of 2×10^{-4} T as shown. What is the magnitude and direction of the magnetic force on the wire if it is 2 meters long?

Solution: Memorize (or write on your equation sheet) the formula for the magnetic force on a wire:

$$F = BIL\sin\theta = 2\times10^{-4}\,\text{T}(15\text{A})(2\text{m})(1) = 6\times10^{-3}\,\text{N}$$

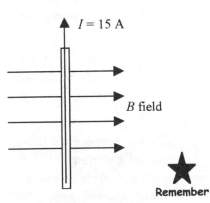

$I = 15$ A

B field

Remember

To find the direction of the force place your right hand flat down, fingers to the right along the magnetic field and thumb up. The direction of the current is the same as the direction of a single positive charge. The current is just a collection of charges. The force is into the page.

18-5 If the wire in the problem above were tilted at an angle of $30°$ with respect to the vertical as shown, calculate the new force.

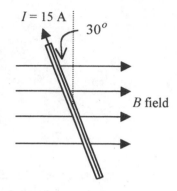

$I = 15$ A

$30°$

B field

Watch Out!

Solution: Be careful. The angle specified in the problem here is not the same as the angle θ in the formula. The angle θ in the formula refers to the angle between the magnetic field and the wire which for this problem is $60°$.
The answer to the question is just the previous answer times $\sin 60° = 0.866$. The answer is 5.2×10^{-3} N.

18-6 A loop of wire is perpendicular to a magnetic field of 3×10^{-3} T. It carries a current of 3 A and has a radius of 100 cm. What happens to the loop?

Solution: The loop experiences a **torque** rather than a force. The formula is:

$\tau = BIA \sin \theta$, where A is the area of the loop. $A = \pi (1m)^2 = 3.14 m^2$. Notice we are always using meters for length, area etc.

$$\tau = BIA \sin \theta = (3 \times 10^{-3} \text{ T})(3\text{A})(3.14 m^2)(1) = 0.0283 \text{ N} \cdot \text{m}$$

Remember that the units of torque are $\text{N} \cdot \text{m}$ (and it is not written as Joules which would be the same units, but Joules is for energy). Torque was presented in the first semester of physics. Forces move things in a straight line, torques spin things around on an axis.

18-7 Two long parallel wires separated by 0.60 m carry antiparallel currents (the currents are physically parallel but in opposite directions) of 7.0 A each. What is the resulting field along the line between the wires and 0.40 m from the lower wire?

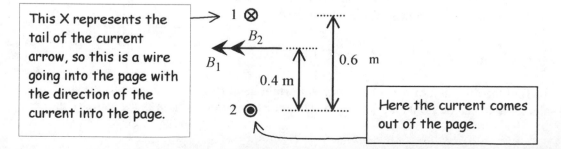

This X represents the tail of the current arrow, so this is a wire going into the page with the direction of the current into the page.

1 ⊗

B_2

B_1

0.6 m

0.4 m

2 ◉

Here the current comes out of the page.

Solution: At the specified position the magnetic field due to wire 1 is

$$B_1 = \frac{\mu_o}{2\pi}\frac{I}{r} = \frac{4\pi \times 10^{-7}\,\text{Wb}/\text{A}\cdot\text{m}(7.0\text{A})}{2\pi \cdot 0.20\,\text{m}} = 7.0 \times 10^{-6}\,\text{T}$$

The way to find the direction of a magnetic field produced by a current carrying wire is shown in the illustration. The magnetic field curls all the way around the wire. It makes a complete circle around the wire. The strength of the field falls off as $1/r$.

Thumb of right hand in the direction of the current.

Fingers curl around the wire in the direction of the magnetic field.

For our situation, the magnetic field for the top wire is directed to the left of the page. We find this by pointing the thumb of our right hand into the page. Curling the fingers says that the magnetic field makes a clockwise-direction circle around the wire.

The magnetic field due to wire 2 is

$$B_2 = \frac{\mu_o}{2\pi}\frac{I}{r} = \frac{4\pi \times 10^{-7}\,\text{Wb}/\text{A}\cdot\text{m}(7.0\text{A})}{2\pi \cdot 0.40\,\text{m}} = 3.5 \times 10^{-6}\,\text{T}$$

This time we place the thumb out of the page, with fingers curling in the counter-clockwise direction. But we are looking at a point on the top side of this wire, as opposed to the bottom side of the previous wire. The magnetic field therefore turns out to be in the same direction. The magnitude of total magnetic field is just the sum of the two values we have, or $10.5 \times 10^{-6}\,\text{T}$.

18-8 Calculate the magnetic field $10\,\text{cm}$ perpendicular to the bisector of the line connecting two wires $2.0\,\text{cm}$ apart and carrying antiparallel currents of $100\,\text{A}$ each as shown.

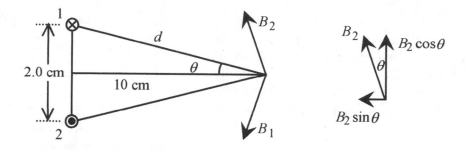

Solution: The magnitude of the magnetic field from each wire is $\mu_o I/2\pi d$. Now, consider the symmetry of the problem. Because of the geometry, the vertical components of B_1 and B_2 add to zero. The horizontal components add to a vector pointing toward the centerline of the wires with magnitude $B = 2 B_1 \sin\theta$. The dimension $d = \sqrt{10^2 + 1^2} = \sqrt{101}\,\text{cm}$. The $\sin\theta = 1.0/\sqrt{101}$. The magnitude of the resultant field is

$$B = 2\times\frac{\mu_o I}{2\pi d}\sin\theta = \frac{2\cdot 4\pi\times 10^{-7}\,\text{T}\cdot\text{m}/\text{A}(100\,\text{A})}{2\pi\sqrt{101}\times 10^{-2}\,\text{m}}\frac{1.0}{\sqrt{101}} = 4.0\times 10^{-5}\,\text{T}$$

18-9 Consider two wires coming out of the page carrying equal, and oppositely directed, currents as shown at the top of the next page. The wires are suspended from a common point with $0.40\,\text{m}$ long cords. The wires have a mass per unit length of $30\times 10^{-3}\,\text{kg/m}$. Calculate the current to produce an angle of 6°.

Solution: This is a force-balance problem. The horizontal component of the tension (force) in the cord must equal the magnetic force due to the currents. The magnetic force due to the currents pushes the wires apart. Look at $1.0\,\text{m}$ of the wire. The mass of $1.0\,\text{m}$ of wire is $30\times 10^{-3}\,\text{kg}$, and the force due to gravity is $mg = 0.29\,\text{N}$. This must equal the vertical component of the tension in the cord. From the geometry, we calculate

$$\tan 6^o = F_h/F_v \quad\text{or}\quad F_h = 0.29\,\text{N}(\tan 6^\circ) = 0.031\text{N}$$

In the following figure, the separation of the wires is $2s$ where s is defined in the figure through $\sin\quad 6^o = s/0.40\,\text{m}$ or $s = 0.40\,\text{m}\cdot\sin 6^o$.

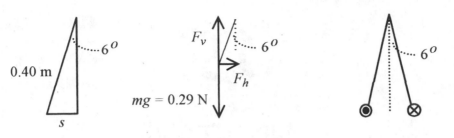

The electric force produced by the current in one wire producing a magnetic field that acts on the current in the other wire is

$$F = BI = \frac{\mu_0 I^2}{2\pi(2s)}$$

and this force must equal the mechanical force, or tension, in the cord, so

$$0.031\,\mathrm{N} = \frac{4\pi \times 10^{-7}\,\mathrm{T \cdot m/A} \cdot I^2}{2\pi(2 \cdot 0.40\,\mathrm{m} \cdot \sin 6^o)}$$

Solving for I, we get

$$I = \left[\frac{0.40 \cdot \sin 6^o (0.031)}{10^{-7}} \mathrm{A}^2 \right]^{1/2} = 114\,\mathrm{A}$$

18-10 Enough of this stuff. It's time to eat. For this we use the right hand rule as shown below.

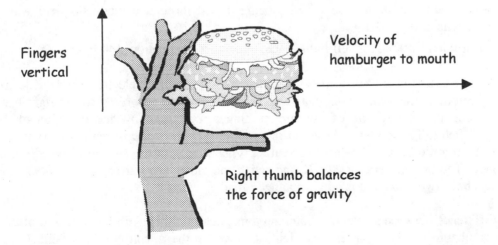

Be careful not to spill ketchup on the page – you may blot out a portion of a formula and end up copying it down wrong.

Watch
Out!

19

INDUCTION

This is an important chapter to understand **conceptually**. It may at first seem difficult to understand some of the concepts in this chapter, but induction is very important in the world we live in. The principles in this chapter govern how we generate electric power. Recording and playback of audio tapes is done with the principle of magnetic induction. Microphones, electric guitars, and other musical instruments use induction. Electric motors and electrical transformers operate by induction. There are many more devices that could be used as examples. Most texts have explanations of how these devices operate. When you have figured out some of the basic concepts, go back and read the sections of the text that cover these devices. You should find it interesting.

Things to Remember

- Magnetic flux is given by: $\Phi = B_\perp (Area)$. By $B_\perp$, we mean the component of the magnetic field that is perpendicular to the loop. The unit of flux is Webers.

- Induced *emf* through a coil of wire is $\mathcal{E} = -N\dfrac{\Delta\Phi}{\Delta t}$. N is the number of turns of wire.

- Induced *emf* due to a loop of wire being pulled through a magnetic field with a velocity is *emf* $= -BLv$.

- For problems asking for self inductance, remember the relation $N\Phi = LI$.

OK, let's start simple. **Stationary** electric charges produce electric fields. Electric currents produce magnetic fields that wrap around the wires according to the right hand rule (thumb in the direction of the current, fingers curling, show the direction of the magnetic field). The top wire shown in the middle section of the figure below (with the dot in the middle of it) indicates a current carrying wire with the current directed out of the page. The bottom wire has an X, indicating the tail of the current arrow. Stationary magnets have two poles and produce magnetic fields.

The difference between an electric field and a magnetic field is as follows. If you place a positive charge **at rest** in an electric field, it moves in the direction of the field. If you place a positive charge **at rest** in a magnetic field, it moves **nowhere**. Only a moving electric charge that has a component of its velocity **perpendicular** to a magnetic field

134

experiences a force. Is the force in the direction of the magnetic field? No. Is the force in the direction of the charge's velocity? No. The force is perpendicular to both and is given by the right-hand rule (fingers = field, thumb = velocity, palm = force).

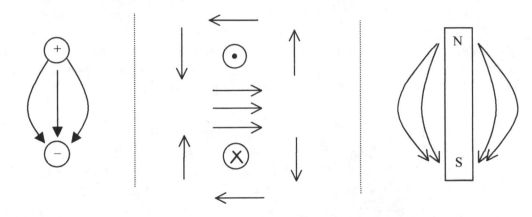

Moving electric charges produce magnetic fields and only moving electric charges are effected by magnetic fields. What do you suppose is inside a stationary magnet then? If you said, "moving electric charges," then you are ready to move on to the bonus round. You may be having a tough time picturing moving electric charges inside a piece of metal, but remember that electrons move in circles around the nucleus of all the atoms in the metal. In non-magnetic materials, all the little magnetic fields produced by these electrons cancel out. In a magnet some of the little fields add together, so you get a macroscopic field.

Faraday's Law

There are two basic experiments demonstrating Faraday's law. The first experiment involves passing a magnet through a loop of wire. If a magnet is passed through a loop of wire connected to a galvanometer as shown below, then three things are observed:

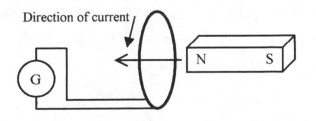

1) A current is observed in the loop when the magnet is moving. If the magnet is still, you get nothing (sound familiar?).
2) The direction of the current depends on the direction of the magnet. The direction of the current is shown above for the case where the magnet is moving

as shown (to the left). If the magnet moves the other way, then the direction of the current is reversed.

3) The magnitude of the deflection (current) is proportional to (a) the strength of the magnet and (b) its velocity.

19-1 Let's put in some numbers for the above situation. The way we would say it is this: The magnetic field goes from 0 Teslas to a strength of 2 Teslas in 5 seconds. The coil has a radius of 20 cm. Find the average induced *emf* in this time.

Solution:

First let's handle that new buzz word: **emf**. It stands for electromotive force. It's a fancy way of saying **voltage**.

Thank you. To find the induced *emf*, we use the formula: $\varepsilon = -N\dfrac{\Delta\Phi}{\Delta t}$. N is the number of turns of wire. For this case, it is just one. $\Delta\Phi$ is the change in magnetic flux. Flux is magnetic field perpendicular to the coil times area. So if the wire had a smaller radius, then the magnetic flux through the middle would be smaller and the induced voltage would be smaller, even though the magnetic field strength is the same. Putting the numbers in:

$$\varepsilon = -\frac{N(B)(Area)}{\Delta t} = -\frac{1(2\,\text{T})\,\pi(0.2\,\text{m})^2}{5\,\text{s}} = -0.05\ \text{volts}$$

If we moved the magnet faster, say taking only 0.5 second, then the induced voltage would be ten times higher, or 0.5 volt.

19-2 A 20-turn coil of radius $0.050\,\text{m}$ is placed in a position where the magnetic field is changed from $40\times10^{-3}\,\text{T}$ in one direction to $40\times10^{-3}\,\text{T}$ in the other direction in a time of $60\times10^{-3}\,\text{s}$. Find the induced *emf*.

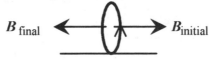

$B_{\text{final}} \longleftarrow \bigoplus \longrightarrow B_{\text{initial}}$

Solution: Inside the coil, the flux is $\Phi = B\ \ A = 40\times10^{-3}\,\text{T}\,\pi(0.05\,\text{m})^2 = 3.1\times10^{-4}\,\text{Wb}$. Wb stands for Webers, the unit of flux. The change in flux is $\Delta\Phi = 6.2\times10^{-4}\,\text{Wb}$, and this occurs over $60\times10^{-3}\,\text{s}$ in the coil of 20 turns, so the induced *emf* is

$$emf = -N\frac{\Delta\Phi}{\Delta t} = -20\frac{6.2\times10^{-4}\,\text{Wb}}{60\times10^{-3}\,\text{s}} = -0.21\text{V}$$

So far we have left out mentioning one thing: that minus sign in the formula for *emf*. Let's get to what this means by describing another experiment.

The second Faraday experiment involves two loops of wire. If the current in the primary loop (the one with the battery) is **changed**, then there is a current in the secondary loop (the one with the galvanometer).

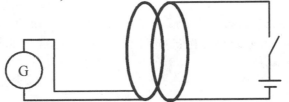

If the switch connecting the battery to the loop is opened and closed, three things are observed:

1) A current in the galvanometer loop is observed when the switch is closed and the current goes from not flowing at all to flowing.
2) A current in the opposite direction is observed when the switch is opened and the current goes from flowing to not flowing.
3) No current in the galvanometer loop is observed when there is a steady current and no current is observed when there is zero current in the battery loop (sound familiar?).

An *emf* and a current is induced in a loop when there is a **change** in magnetic field in the loop. The magnitude of the induced *emf* is proportional to the rate of change of flux. Now here's the clincher: **The induced *emf* in turn sets up a current that itself produces a magnetic field to oppose** *the change* **of the applied magnetic field**. Yes, this is a mouthful, but it is important that you eventually "get" this. It might take a while, but it's worth it. Notice, we did not say that the induced field opposes the magnetic field. It opposes a change in the magnetic field. If the applied magnetic field is strong and you reduce it, the current induced in the wire will try to reinforce the applied magnetic field. By the way, this is why there is a minus sign in the equation for *emf*. It's just saying that the voltage opposes the change in the field so we put a minus sign in the equation. Let's not worry any more about that minus sign.

Lenz's Law

To determine the direction of the induced current, look at the coil and the direction of the initial and final magnetic fields back in problem 19-2. The field, originally pointing to the right, collapses to zero and then grows to the left to its final value. The current is physically constrained by the wire to go only in one of two directions. If the current in the coil were such that the field produced by this current grew in the same direction as the field that initiated the current, then the current would continue to grow (because the field would continue to change in the same direction). This would mean that any time a field changed in a loop of wire, the current in the loop would grow on and on forever without limit. This is clearly contrary to nature. **Lenz's law,** applied to this situation, states that an induced current is in a direction so as to produce a field that opposes the changing field that initiated the current. This will be shown in subsequent problems.

19-3 Take a 50-turn rectangular coil of dimensions $0.10\,\mathrm{m}$ by $0.20\,\mathrm{m}$ and rotate it from a position perpendicular to a field of $0.50\,\mathrm{T}$ to parallel to the field in $0.10\,\mathrm{s}$ and calculate the induced *emf*.

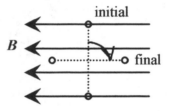

Solution: The flux $\Phi = BA$. In addition, $\Delta\Phi = BA$ since the flux goes from maximum to zero.

$$\Delta\Phi = BA = 0.50\,\mathrm{T} \cdot 2.0\times10^{-2}\,\mathrm{m}^2 = 1.0\times10^{-2}\,\mathrm{T}\cdot\mathrm{m}^2$$

The coil is 50 turns, so the induced *emf* is

$$\textit{emf} = -N\frac{\Delta\Phi}{\Delta t} = -50\frac{1.0\times10^{-2}\,\mathrm{T}\cdot\mathrm{m}^2}{0.10\,\mathrm{s}} = -5.0\,\mathrm{V}$$

19-4 A loop of width $0.20\,\mathrm{m}$, length $0.80\,\mathrm{m}$, and resistance $200\,\Omega$ is pulled through a magnetic field of $0.40\,\mathrm{T}$ at $0.20\,\mathrm{m/s}$. (a) Find the induced *emf* and current. (b) Find the force necessary to pull the loop. (c) Find the work performed, and the power, the rate of doing the work.

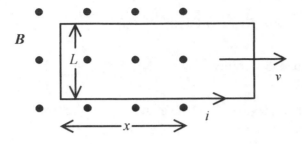

The total flux within the loop is $\Phi = BLx$.

The induced *emf* is

$$emf = -\frac{\Delta\Phi}{\Delta t} = -BL\frac{\Delta x}{\Delta t} = -BLv.$$

As the loop is being pulled through the field, there is a decrease in the number of lines of magnetic flux pointing out of the page. Therefore the induced current will be in a direction so as to create more flux out of the page. This requires a counterclockwise current around the loop. Interestingly, this current produces forces on the sides of the loop. Apply the right hand rule to the forces and note that the force on the bottom section of the wire is down, while the force on the top section of the wire is up. These two forces are equal and opposite and so add to zero. The force on the vertical section of the wire is to the left in opposition to the force producing the velocity. The force on the wire due to the induced current is BiL. The work performed by this force is (in analog with mechanics) force times distance or $F \cdot x = BiLx$. The power (again, in analog with mechanics) is force times velocity or $P = F \cdot v = BiLv$.

Solution: (a) The induced *emf* is

$$emf = -BLv = -0.40\,\text{T} \cdot 0.20\,\text{m} \cdot 0.20\,\text{m/s} = -1.6\times10^{-2}\,\text{V}.$$

The induced current (in the direction shown in the figure above) is

$$i = \frac{emf}{R} = \frac{1.6\times10^{-2}\,\text{V}}{200\,\Omega} = 8.0\times10^{-5}\,\text{A}$$

(b) The force necessary to pull the wire out of the field is

$$F = BiL = 0.40\,\text{T} \cdot 8.0\times10^{-5}\,\text{A} \cdot 0.20\,\text{m} = 6.4\times10^{-6}\,\text{N}$$

(c) The work performed in completely removing the loop is this force times the length of the loop. Remember that the basic definition of work is $\text{force} \times \text{distance}$.

$$W = F \cdot x = 6.4 \times 10^{-6} \, N \cdot 0.80 \, m = 5.1 \times 10^{-6} \, J$$

The power delivered to the loop is $P = F \cdot v = 6.4 \times 10^{-6} \, N \cdot 0.20 \, m/s = 1.3 \times 10^{-6} \, W$.

19-5 A variation of the problem done above is one where the loop is replaced by a U-shaped piece of wire with a sliding piece along the arms of the "U." Take a $0.60 \, T$ field and a U- shaped piece with width $0.30 \, m$. The entire circuit has resistance of $20 \, \Omega$ and the sliding bar is moving to the right at $6.0 \, m/s$.

Solution: The *emf* generated in the wire and moving rod is

$$emf = -B \quad L \quad v = -0.60 \, T \cdot 0.30 \, m \cdot 6.0 \, m/s = -1.1 \, V$$

The current in the loop is $i = emf/R = 1.1 \, V/20\Omega = 0.054 \, A$.

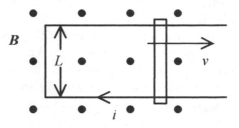

When the bar is moving to the right, the number of lines of flux (amount) is increasing out of the page, so the current is clockwise so as to produce a field pointing into the page, i.e., a field opposing the increasing field, within the "loop" causing the current.

The work performed in moving the bar is

$$W = Fv = BiLv = 0.60 \, T \cdot 0.054 \, A \cdot 0.30 \, m \cdot 6.0 \, m/s = 0.058 \, J$$

Self-Inductance

A changing current in an isolated coil produces an *emf* in itself. This is called a self-induced *emf*. The direction of the self-induced *emf* is so as to produce a current that opposes the current that set it up.

Calculation of the self-induced *emf* follows from a consideration of the geometry of the coil. Consider a close-packed coil with a flux, Φ, passing through each of N turns of the coil. The number of turns times the flux is called the flux linkages, $N\Phi$. The self-induced *emf* is proportional to the time rate of change of these flux linkages.

19-6 For a coil of 300 turns and inductance 12×10^{-3} H, find the flux through the coil when the current is 5.0×10^{-3} A.

Solution: Use the relationship between flux linkages and inductance $N\Phi = LI$.

$$\Phi = \frac{LI}{N} = \frac{12 \times 10^{-3}\,\text{H} \cdot 5.0 \times 10^{-3}\,\text{A}}{300} = 2.0 \times 10^{-7}\,\text{Wb}$$

Mutual Inductance

If there are coils within, or adjacent to, other coils, then we can define mutual inductance, the linking of one coil with another via the flux. If the flux is changing, then the current in the one coil is linked to the current in the other coil.

19-7 Consider a coil within a coil. The larger coil is 3.0 cm in radius and we are looking at a cross section in the diagram below. The coil is 10 cm in length with 1200 turns. The smaller coil is 2.0 cm in radius, 4.0 cm in length with 50 turns. Its wires are perpendicular to the first coil. (a) Calculate the inductance. (b) Find the voltage induced in the second coil as a result of a steady change in current in the first coil of 0.50 A/s .

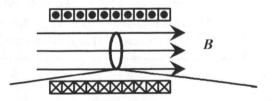

Solution: (a) The current in the large solenoid sets up a magnetic field within the coil of $B = \mu_0 Ni/\ell$. Assume the flux is uniform across the inside of the large coil so that the flux through the small coil is this B times the area of the small coil $\Phi_2 = BA_2$. In this problem we will use the subscript 1 to refer to the large coil and 2 to refer to the smaller coil. The mutual inductance, in accord with the definition of self-inductance, is the number of flux linkages divided by the current.

$$L = \frac{N_2 \Phi_2}{i_1} = \frac{N_2}{i_1} BA_2 = \frac{N_2}{i_1} \frac{\mu_0 N_1 i_1}{\ell} A_2 = \mu_0 \frac{N_1}{\ell} N_2 A_2$$

The inductance is totally geometry dependent. For this particular combination of coils we calculate the following:

$$L = \mu_o \frac{N_1}{\ell} N_2 A_2 = 4\pi \times 10^{-7} \,\text{T} \cdot \text{m/A} \frac{1200}{0.10 \,\text{m}} (50)\pi(4.0 \times 10^{-4} \text{m}^2) = 9.5 \times 10^{-4} \text{H}$$

(b) The voltage induced in the second coil as a result of a steady change in current of $0.50 \,\text{A/s}$ is calculated as follows:

$$\textbf{\textit{emf}} = L \frac{\Delta i}{\Delta t} = 9.5 \times 10^{-4} \,\text{H} \frac{0.50 \,\text{A}}{\text{s}} = 4.7 \times 10^{-4} \,\text{V}$$

There are many different kinds of mutual inductance problems. Watch for these buzzzz words: **transformer, primary coil and secondary coil, generator.** You don't have to have one coil inside another for a mutual inductance problem; they could be next to each other. The main thing to remember is that the magnetic flux from one coil induces an **emf** in the other.

Also, **solenoids** and **toroids** are different shapes of coils of wire. If you see a question asking for the inductance of a solenoid, it probably is asking for the self inductance of this coil of wire, so use $N\Phi = LI$.

20

AC CIRCUITS

Things to Remember

- Ohm's Law, $V = IR$, is the general equation for a dc (direct current) circuit. For ac (alternating current) circuits the equation is $V = IZ$, where Z is the impedance of the circuit (also measured in Ohms). Impedance just means resistance to alternating current.
- Impedance consists of (1) resistance from a resistor, (2) reactance from a capacitor (X_C), and (3) reactance from an inductor (X_L). The equation for impedance is

  $Z = \sqrt{R^2 + (X_L - X_C)^2}$, where $X_C = \dfrac{1}{2\pi f C}$ and $X_L = 2\pi f L$.

- The phase angle between current and voltage in an ac circuit is given by:

  $\tan\phi = \dfrac{X_L - X_C}{R}$.

- The root-mean-square voltage is related to the maximum voltage in an ac circuit by:

  $V_{rms} = \dfrac{V_{max}}{\sqrt{2}}$.

- The values of $V/\sqrt{2}$ and $I/\sqrt{2}$ are used to compute the average power dissipated in an ac circuit and are equivalent to V and I used to compute power in a dc circuit. dc voltmeters and ammeters measure V and I with the product being power, $P = VI$. Alternating current voltmeters and ammeters must measure $V/\sqrt{2}$ and $I/\sqrt{2}$, the time average of these quantities, so power calculations in ac and dc will be the same. The two formulas to remember for power dissipation in an ac circuit are: $P = VI\cos\phi$ and

  $P = I^2 R$.

Analysis of Circuit Elements

Lets talk about resistors, capacitors, and inductors in a dc circuit for a moment. Suppose you have a battery hooked up to a resistor with a switch in the circuit as shown in the figure on the next page. When the switch is closed, you immediately get 10 volts across the resistor. The resistor always puts up a resistance of 10 Ohms. The current that flows in the circuit is given by $V = IR$. So for this case we should get 1 amp.

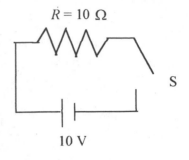

Now suppose you have an inductor hooked up instead of the resistor. The switch connects the battery to the inductor. What happens? Well, again a voltage appears across the inductor and current begins to flow. As soon as current starts to flow, the inductor becomes resistive. An inductor is a coil of wire with self inductance that we saw in the last chapter. The inductor opposes **change** in the current. So if there is originally no current and we try to turn current on, the inductor acts like a resistor. The inductor is doomed to fight a loosing battle, however. Eventually the current will continue to increase and then we will get a steady current flow through the inductor. Eventually the current flow in this circuit is limited only by the resistance of the wire. Now suppose that we open the switch. The current tries to drop to zero since the voltage is removed. The inductor now makes a last ditch effort to try to keep the current going. Remember, it doesn't like changes in current. The inductor has gotten used to current flowing through it. Now that the current is decreasing, the inductor tries to keep it going. But again, the inductor is not a power source, so the current eventually does drop to zero, but more slowly than it would have if no inductor had been present.

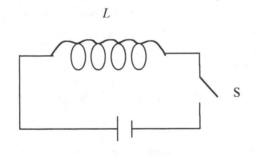

Now let's put a capacitor in the dc circuit and throw the switch. A voltage appears across the plates of the capacitor, current flows and the plates begin to accumulate charge. In the early stages of the current flow, the capacitor doesn't mind much. It doesn't put up much resistance in the beginning. But after a while, lots of charge begins to build up on the plates of the capacitor. Positive charge builds up on the positive plate and negative charge builds up on the negative plate. Now if you have a plate full of positive charge and you try to keep putting more and more positive charge on, what do you think happens? Like charges repel. The more charge that is on the capacitor, the greater the resistance of the capacitor. Eventually, the capacitor reaches capacity (sorry for the pun, but that's why it is called a capacitor). It can't hold any more charge. The resistance is essentially infinite and the current flow goes to zero.

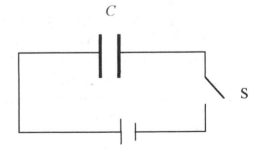

Now, notice this about the inductor vs. the capacitor. When we first threw the switch to start the current flow on the inductor, it immediately put up a resistance. Toward the end, when lots of current was flowing and the current flow was limited only by the resistance of the wire, we had huge current flow and the inductor had given up trying to resist because the large current flow was constant. The capacitor was the other way around. It didn't mind in the beginning when we first turned on the current, but later on it put up huge resistance to the point where it put up infinite resistance. So it seems like the inductor and the capacitor are opposites in some sense. To put it more technically in the language of ac circuits, they are 180° out of **phase** with each other.

How do each of these 3 circuit elements behave in an alternating current circuit? Having these 3 elements in an ac circuit is like having a battery that is constantly reversing its polarity. First the current flows one direction, say clockwise around the cicuit, then counter-clockwise, then back to clockwise, and so forth. Instead of resistance, we have reactance of each individual element and an overall impedance for the entire circuit. The resistor is easy – it always has the same resistance. The inductor and capacitor are not so easy.

20-1 A $20\,\mu F$ capacitor is connected to a variable frequency ac source with maximum voltage $30\,V$. What is the capacitive reactance at 60 Hz, 600 Hz and 60 kHz?

Solution: $X_C|_{60} = \dfrac{1}{\omega C} = \dfrac{10^6\,s}{2\pi \cdot 60 \cdot 20\,F} = 130\,\Omega \quad X_C|_{600} = 13\,\Omega \quad X_C|_{60k} = 0.13\,\Omega$

Remember the capacitor gets resistive (reactive) if you put a lot of charge on its plates. If you change the direction of the current fast enough, the capacitor doesn't react as much because there is not enough time to build up a lot of charge on the plates.

Insight

20-2 A $120\,mH$ inductor is connected to a variable frequency ac source with maximum voltage $10\,V$. What is the inductive reactance at 100 Hz and 1.0 MHz?

Solution: $\qquad X_L|_{100} = \omega L = 2\pi \cdot 100\,Hz \cdot 0.12\,H = 75\,\Omega$

$\qquad\qquad X_L|_{1.0M} = 2\pi \cdot 1.0 \times 10^6\,Hz \cdot 0.12\,H = 7.5 \times 10^5\,\Omega$

Here we see that the inductor gets more reactive at higher frequency. Remember that the inductor opposes change in current. The faster we try to change the current, the more reactive the inductor becomes.

Insight

20-3 At what frequency do a $65\,mH$ inductor and a $20\,\mu F$ capacitor have the same reactance?

Solution: When $X_L = X_C$, or $\omega L = \dfrac{1}{\omega C}$. The frequency then is

$$\omega = \frac{1}{\sqrt{LC}} \quad \text{or} \quad f = \frac{1}{2\pi\sqrt{LC}}$$

This is the condition for **resonance** or **oscillation** in the L-C circuit.

$$f = \frac{1}{2\pi\sqrt{65\times10^{-3}\,\text{H}\cdot20\times10^{-6}\,\text{F}}} = 140\,\text{Hz}$$

20-4 An ac source of $100\,\text{Hz}$ and maximum voltage of $20\,\text{V}$ is connected to a $70\,\text{mH}$ inductor. What is the maximum current? When the current is maximum, what is the voltage of the source?

Solution: First calculate the inductive reactance as follows:

$$X_L = \omega L = 2\pi\cdot100\,\text{Hz}\cdot0.070\,\text{H} = 44\,\Omega$$

The maximum current is $I_L\big|_{max} = V_L/X_L = 20\,\text{V}/44\,\Omega = 0.45\,\text{A}$

20-5 An R-L-C circuit with $R = 200\,\Omega$, $L = 0.40\,\text{H}$, and $C = 3.0\,\mu\text{F}$ is driven by an ac source of $20\,\text{V}$ maximum and frequency $100\,\text{Hz}$. Find the maximum current and voltages across each of the components.

Pattern

Solution: R-L-C circuit problems can be confusing. The key to successfully solving them is to follow a logical path through the problem. The current is everywhere the same and is determined by the source voltage and the impedance. The impedance is determined by the resistance and the reactances, and the reactances are frequency dependent. As you proceed through this problem be aware of the logic in the calculations.

1. Find the reactances.

$$X_L = \omega L = 2\pi\cdot100\,\text{Hz}\cdot0.40\,\text{H} = 251\,\Omega$$

$$X_C = \frac{1}{\omega C} = \frac{10^6}{2\pi\cdot100\,\text{Hz}\cdot3.0\,\text{F}} = 530\,\Omega$$

$$(X_L - X_C)^2 = (251\,\Omega - 530\,\Omega)^2 = 77{,}800\,\Omega^2$$

2. Find the impedance.

$$Z = \sqrt{R^2 + (X_L - X_C)^2} = \sqrt{(200)^2 + 77800}\ \Omega = 343\ \Omega$$

3. Find the current.

$$I = V/Z = 20\text{V}/343\ \Omega = 0.058\ \text{A}$$

4. Find the voltages across each element.

$$V_R = 0.058\ \text{A} \cdot 200\ \Omega = 11.6\ \text{V}$$

$$V_L = 0.058\ \text{A} \cdot 251\ \Omega = 14.6\ \text{V}$$

$$V_C = 0.058\ \text{A} \cdot 530\ \Omega = 30.7\ \text{V}$$

You may notice here that the addition of the voltages across the three circuit elements do not add up to the source voltage. Also, it is interesting to note that the voltage across the capacitor is greater than the source voltage. When you understand phasors, you will understand why $V_R + V_L + V_C$ does not add up to the source voltage.

Phasors

Find out if you are going to need to know phasor diagrams for your next test. Some instructors include these types of problems, and some don't. This section on phasors is tricky so don't spend a lot of time on it if this won't be on your test.

Problem 20-5 is the easier version of an R-L-C circuit problem, but there is a harder version that also asks for the **phase** between the current and the voltage. Remember that for the above problem, we just found the maximum voltage across each circuit element. This means that during the cycling of the ac voltage, these are the highest values across each element. But what we haven't done is found the phase relationships between the voltages. In other words, the maximum voltage across the capacitor doesn't occur at the same time as the maximum voltage across the inductor.

The phasor diagrams are most helpful in understanding R-L-C circuits. An R-L-C circuit means that the resistor, capacitor, and inductor are wired in series. There are two important points to keep in mind in the analysis of these circuits. First, the sum of the instantaneous voltages must equal the source voltage: $V\cos\omega t = v_R + v_L + v_C$. Second, since there is only one current path, the current is everywhere the same. Voltages on the various components have different phase relationships, but the current is the same everywhere in the circuit. The phasor diagram for a typical R-L-C circuit is shown below. Do not try to take this in all at once. Follow along the steps in the construction of the diagram.

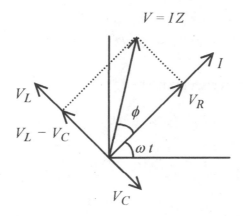

Pattern

Place the I phasor at the arbitrary angle ωt.

Place the V_R phasor over I. The voltage and current in the resistor are in phase.

Add $V_L = I X_L$ leading V_R by 90°.

Add $V_C = I X_C$ lagging V_R by 90°.

On an axis perpendicular to V_R and I, V_L and V_C, add in a vector manner to produce $V_L - V_C$. In this example, $V_L > V_C$.

If the load is resistive, voltage is in phase with current. If the load is entirely inductive or entirely capacitive the voltage is 90° out of phase with current. In this situation, with all elements present, the voltage is the vector-like sum of V_R and $V_L - V_C$.

The phase relation between V and I is seen from the phasor diagram as

★

Remember

$$\tan\phi = \frac{V_L - V_C}{V_R}$$

To obtain a better picture of what is going on here, imagine measuring the ac voltages of the source, resistor, capacitor, and inductor, and the current in the circuit. The voltages across the resistor, capacitor, and inductor do not add up to the source voltage! They are not in phase! The source voltage divided by the impedance will equal the current. Finally the phase angle between the source voltage and current comes from the equation above.

20-6 An R-L-C circuit with $R = 200\,\Omega$, $L = 0.40\,H$, and $C = 3.0\,\mu F$ is driven by an ac source of 20 V maximum and frequency 100 Hz. This is the same circuit that we used in the last problem. Construct the phasor diagram and find the angle of the source voltage with respect to the current.

Solution: The phasor diagram is shown below. Draw the phasor representing $I = 0.058\,A$ at an arbitrary angle ωt. Draw the phasor representing $V_R = 11.6\,V$ in the same direction as I. Draw $V_L = 14.6\,V$ leading V_R by 90°. Add $V_C = 30.7\,V$ lagging V_R by 90°. Complete the rectangle with sides $V_L - V_C$ and V_R. Since $V_C > V_L$, this vector points in the same direction as V_C. The source voltage V is the diagonal of this rectangle. The phase angle is determined by

$$\tan\phi = \frac{V_L - V_C}{V_R} = \frac{14.6 - 30.7}{11.6} \quad \text{or} \quad \phi = -54^o$$

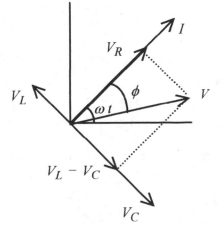

Because this angle is negative, the load is resistive and capacitive and (from the diagram) the source voltage lags the current by 54°.

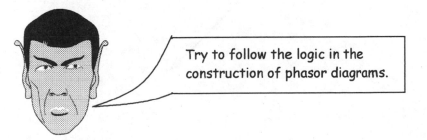

Try to follow the logic in the construction of phasor diagrams.

After you are clear on the calculations go back over these last two problems and concentrate on the logic. The biggest pitfall in R-L-C circuit problems is losing your way!

20-7 An R-L-C circuit consists of a $300\,\Omega$ resistor, $0.15\,H$ inductor, and $4.5\,\mu F$ capacitor driven by an ac source of 800 rad/s (or Hz). Find the phase angle of the source voltage with respect to the current.

Solution: Notice that the maximum voltage is not given in the problem, yet a phase diagram is to be drawn. The problem is written this way to emphasize that it is not necessary to know the maximum source voltage to find the phase angle. The phase diagrams so far have used voltage. The voltage across each component is the resistance or reactance times a constant, the current. Thus the diagram can be drawn with resistance and reactances instead of voltages.

Calculate the reactances: $X_L = \omega L = 800\,\mathrm{rad/s} \cdot 0.15\,\mathrm{H} = 120\,\Omega$

$$X_C = \frac{1}{\omega C} = \frac{1}{(800\,\mathrm{rad/s})4.5 \times 10^{-6}\,\mathrm{F}} = 278\,\Omega$$

Now draw the phasor diagram as shown. Draw I at some arbitrary angle ωt. Draw R along I proportional to $300\,\Omega$. Draw X_L leading R by 90° proportional to $120\,\Omega$. Draw X_C lagging R by 90° proportional to $278\,\Omega$. Draw $X_L - X_C = -158\,\Omega$. Complete the rectangle formed by R and $X_L - X_C$ and draw the diagonal, which is proportional to Z.

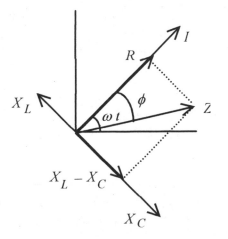

The phase angle is

$$\tan\phi = \frac{X_L - X_C}{R} = \frac{-158}{300} \quad \text{or} \quad \phi = -28^o$$

The load is resistive and capacitive, and the phase angle is -28° with the source voltage lagging the current.

RMS Values and Power

Let's take a break from the insanity of all these phasor diagrams and do a few more of the easier types of problems that you might encounter.

20-8 An rms voltage of 50 V is applied to an ac circuit with a resistance of 10 Ω. Find (a) the maximum voltage, (b) the maximum current, and (c) the rms current.

Solution: (a) Memorize the conversion for rms to max voltage and current.

$$V_{rms} = \frac{V_{max}}{\sqrt{2}} \Rightarrow 50\,\text{V} = \frac{V_{max}}{1.41} \Rightarrow V_{max} = 70.7 \text{ V}$$

Remember

(b) Ohm's law still works for an ac circuit: $V_{max} = I_{max}R$. So $I = 7.07\,\text{A}$.

(c) Ohm's law also works for rms values: $V_{rms} = I_{rms}R$. So $I = 5\,\text{A}$.

20-9 A series R-L-C circuit has a resistance of 10 Ω and an impedance of 15 Ω. The maximum voltage delivered to the circuit is 50V. (a) What is the average power dissipated in the circuit? (b) What is the phase angle between voltage and current?

Solution: (a) First of all, whenever we talk about average power, we must use rms values for current and voltage. The rms voltage is then $50/\sqrt{2} = 35.4$ V . Even though the circuit has impedance, which means that there is a resistor and/or capacitor present, the power dissipated in an ac circuit can only be due to the resistor. Remember that **the inductor and capacitor do not dissipate power**. The easiest formula to use is $P = I^2 R$. We can find the current first with Ohm's Law.

$$V = IR \Rightarrow 35.4\,\text{V} = I(15\,\Omega) \Rightarrow I = 2.36\,\text{A}$$

$$P = (2.36\,\text{A})^2 10\,\Omega = 56 \text{ Watts}$$

(b) Start with $\tan\varphi = \dfrac{(X_L - X_C)}{R}$, but first we need to get $(X_L - X_C)$ by using the impedance:

$$Z = \sqrt{R^2 + (X_L - X_C)^2} \quad \Rightarrow \quad 15\,\Omega = \sqrt{(10\,\Omega)^2 + (X_L - X_C)^2}$$

Get rid of the square root by squaring both sides: $(15\,\Omega)^2 = (10\Omega)^2 + (X_L - X_C)^2$

$$(X_L - X_C)^2 = 225 - 100 = 125 \quad \Rightarrow \quad (X_L - X_C) = \sqrt{125} = 11.2\,\Omega$$

$$\tan\phi = \frac{11.2}{10} \quad \Rightarrow \quad \phi = \tan^{-1}\left(\frac{11.2}{10}\right) = 48.2^o$$

Whether the angle is positive or negative makes no difference in the power calculation ($\cos\theta = \cos(-\theta)$). If $X_L - X_C$ is positive, then this is a positive angle. If $X_L - X_C$ is negative, then this is a negative angle.

We have completed the answer to part (b), but notice that we can use the other formula to verify the power calculation now:

$$P = VI\cos\phi = 35.5 \text{ V}(2.37\text{A})(\cos 48.2^o) = 56\,\text{Watts}$$

21

LIGHT
(waves, reflection and refraction)

Electromagnetic Waves

Things to Remember

- $B = \dfrac{E}{c}$. This relates the magnetic field, electric field, and the speed of light. Remember that the units of electric field are V/m and the units of magnetic field are Tesla.
- The speed of light is going to pop up in lots of equations for the rest of the course, so memorize the value of the speed of light in a vacuum, which is $c = 3 \times 10^8$ m/s.
- Intensity is given by $S = Power / Area$. Most books use S for intensity – why not I? – who knows?
- The speed of light is related to wavelength and frequency by $c = f\lambda$.

A current in a wire produces a magnetic field, and a changing current produces a changing magnetic field. So if we have an ac current in a wire, we have a changing magnetic field in space, but we also know that the changing magnetic field must in turn produce an electric field. These oscillating fields proceed out together as waves. If you were floating on an inflatable raft in a pool and you sensed a sinusoidal up and down motion of the raft, you would say a (water) wave passed by. If you were viewing a small length of stretched rope and the small segment you were viewing executed a sinusoidal up and down motion, you would say a sine wave travelled down the rope. Likewise if you observed a sinusoidally varying electric field accompanied at right angles by a sinusoidally varying magnetic field, you would say an electromagnetic wave passed by.

21-1 A radio signal at a certain point has a measured maximum electric field of 5.0×10^{-3} V/m. What is the maximum magnetic field?

Solution: The magnetic field maximum is at right angles to the electric field and has the following magnitude.

$$B = \frac{E}{c} = \frac{5.0 \times 10^{-3} \text{ V/m}}{3.0 \times 10^8 \text{ m/s}} = 1.7 \times 10^{-11} \text{ T}$$

The relative strengths of the magnetic and electric fileds are determined by the speed of light.

21-2 How long does it take for an electromagnetic wave to travel to the moon, 3.8×10^8 m away?

Solution: $\text{speed} = \dfrac{\text{distance}}{\text{time}} \Rightarrow \text{time} = \dfrac{\text{distance}}{\text{speed}} = \dfrac{3.8 \times 10^8 \text{ m}}{3.0 \times 10^8 \text{ m/s}} = 1.3 \text{ s}$

21-3 How far does light travel in 1.0 nanosecond $(1.0 \times 10^{-9} \text{ s})$?

Solution: $d = c \quad t = (3.0 \times 10^8 \text{ m/s})(1.0 \times 10^{-9} \text{ s}) = 0.30 \text{ m} \approx 1 \text{ ft}$

This is a handy number to remember if you are trying to picture how long it takes light to travel small dimensions in a room.

21-4 Take the intensity of sunlight at the earth's surface as 1000 W/m^2. Assuming 15% collection efficiency, how much energy is collected on a 1.0 m^2 solar panel exposed to this amount of sunlight for 10 hours? Express your answer (a) in Joules, and (b) in $\text{kW} \cdot \text{hr}$.

Solution: The key to getting problems involving energy, power, and intensity correct is to watch the units closely. In this case notice how the units dictate how to make the calculation. 15% of 1000 is 150 W/m^2.

(a) $\qquad E = \dfrac{150 \text{ W}}{\text{m}^2} \dfrac{\text{J/s}}{\text{W}} \cdot 1.0 \text{ m}^2 \cdot 10 \text{ hr} \dfrac{60 \text{ min}}{\text{hr}} \dfrac{60 \text{ s}}{\text{min}} = 5.4 \times 10^6 \text{ J}$

(b) $\qquad E = \dfrac{150 \text{ W}}{\text{m}^2} 1.0 \text{ m}^2 \cdot 10 \text{ hr} = 1.5 \text{ kW} \cdot \text{hr}$

This is a strange looking unit, but it gives us a feel for the amount of energy that can be collected this way. This 1500 $\text{W} \cdot \text{hr}$ would light fifteen 100 W light bulbs for an hour or run a 1.0 kW microwave oven for 1.5 hours.

21-5 Find the frequency of an argon-ion laser given that its wavelength is 514 nm.

Solution: Use the fact that the velocity of light is equal to frequency times wavelength.

$$c = f\lambda \implies f = \frac{3.0\times10^8\,\text{m/s}}{514\times10^{-9}\,\text{m}} = 5.84\times10^{14}\ \text{waves per second or } 5.84\times10^{14}\ \text{Hz}$$

21-6 The intensity of a source of light is 4.3×10^{-3} Watts per square meter at a distance of 2.0 m from the source. (a) Find the power output of the source. (b) What is the intensity at a distance of 4.0 m?

Solution: (a) Below is shown a slice of the sphere in which the light shines. The total surface area of the sphere is $4\pi r^2$. Memorize this formula. In order to find the power, we need to multiply the intensity (at 0 meters) times the surface area (at 2.0 meters) of the sphere. The sphere here is just adding up all possible space that the light occupies at a distance of 2m.

Remember

$$P = 4.3\times10^{-3}\ \text{W}/\text{m}^2 \cdot 4\pi(2.0\text{m})^2 = 0.21\,\text{W}$$

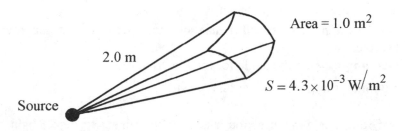

Area = 1.0 m²

2.0 m

$S = 4.3\times10^{-3}\ \text{W}/\text{m}^2$

Source

(b) The same power will exist at the surface of the sphere at 4.0 m.

$$Intensity = Power\,/\,surface\ area \implies S = \frac{0.21\,\text{W}}{4\pi(4\text{m})^2} = 1.04\times10^{-3}\ \text{W/m}^2$$

At twice the distance, the intensity is 4 times less due to the fact that radius is squared in calculating the surface area.

There are a bunch of equations that relate intensity and maximum electric and magnetic field strengths. Find out if you will need these on your test. The problems usually go something like the following:

21-7   What is the maximum electric field strength and intensity of a 500W, spherically symmetric light source at 10 m radius?

Solution: The surface area of a sphere 10m in radius is $A = 4\pi(10\,\text{m})^2 = 1.2 \times 10^3\,\text{m}^2$. The entire 500 W is delivered to this area. The power, intensity, and area are related by

$$S = \frac{P}{A} = \frac{500\,\text{W}}{1.2 \times 10^3\,\text{m}^2} = 0.40\,\text{W/m}^2$$

The electric field intensity comes from $S = \dfrac{E_{max}^2}{2\mu_o c}$

$$E_{max} = \left[2\mu_o cS\right]^{1/2} = \left[2(4\pi \times 10^{-7}\,\text{T·m/A})(3.0 \times 10^8\,\text{m/s})(0.40\,\text{W/m}^2)\right]^{1/2} = 17\,\text{V/m}$$

Reflection and Refraction

Things to Remember

- For reflections, the angle of incidence equals the angle of reflection.
- For refractions, use Snell's Law: $n_1 \sin\theta_1 = n_2 \sin\theta_2$.
- Light slows down in mediums other than air - the speed of light gets divided by the index of refraction.
- The **critical angle** means the angle for total internal reflection and is found by setting $\theta_2 = 90^o$ in Snell's law. Remember that $\sin 90^o = 1$.

In the study of reflection and refraction it is very convenient to depict light as rays. A light ray represents the path of a thin beam (ray) of light.

When light strikes a surface such as an air-glass or air-water interface, part of the incident light is reflected and part is refracted. Reflection of light at any interface follows a very simple law. The angle of incidence equals the angle of reflection with these angles measured from the normal to the surface.

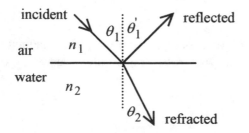

Material	Index of Refraction
vacuum (air)	1.00
glass (crown)	1.62
water	1.33
polystyrene	1.49
diamond	2.42

21-8 A ray of light is incident on a reflecting surface at 70° with respect to the surface. What is the angle of the reflected ray?

Solution: Don't get bagged! Commit it to memory now that anytime we talk about angles with light, it is angles **with respect to the normal**. This means an angle with respect to a line drawn perpendicular to the surface as shown above. The 70° with respect to the surface is 20° with respect to the normal, so the reflected beam is 20° from the normal.

21-9 What is the angle of refraction for a light ray in air incident on glass at an angle of 35° to the normal?

Solution: The angle of refraction is from Snell's law, $n_1 \sin \theta_1 = n_2 \sin \theta_2$. (Refer back to the figure.)

$$1.00 \sin 35^o = 1.62 \sin \theta_2 \quad \text{or} \quad \theta_2 = 21^o$$

The **refracted** (or bent) ray means the one that enters the new medium. The n's here are the indices of refraction. Notice that θ_2 is the alternate interior angle again with respect to the same normal line.

21-10 What is the speed of light in glass?

Solution: Light slows down when it goes into any material other than vacuum. The speed in air and vacuum is essentially the same. The new speed is given by:

$$v = c/n \implies v = \frac{3 \times 10^8 \text{ m/s}}{1.62} = 1.85 \times 10^8 \text{ m/s}$$

21-11 Fiber optic cable is made of light-transmitting material with an index greater than one, so that light injected at the end of the cable will be internally reflected many times over its entire length. Find the critical angle for a fiber optic material with index 1.50.

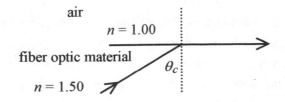

Solution: The critical angle is the one where the angle of refraction is 90°. This means the light will be reflected inside the cable. Applying Snell's law, we calculate

$$1.50 \sin \theta_c = 1.00 \sin 90^o \quad \text{or} \quad \theta_c = 42^o$$

Any beam that strikes the fiber-air interface at an angle greater than 42° will be internally reflected down the fiber. The material that the beam is coming from has to have the higher index of refraction.

21-12 Light is incident at an angle of 35° on a slab of glass 2.0 cm thick. Part of the light travels through the glass and part along its original direction beside the glass. Find (a) the path of the two rays, (b) the velocity of the ray in the glass, (c) and the time difference between the two when they emerge from the opposite side of the glass.

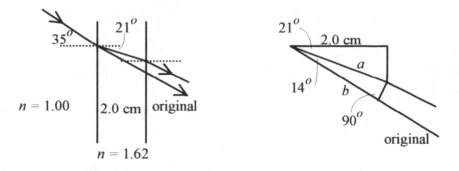

Solution: (a) Apply Snell's law to find the path through the slab.

$$1.00 \sin 35^o = 1.62 \sin \theta_r, \text{ or } \theta_r = 21^o.$$

Because of symmetry, the ray through the slab exits parallel to the original ray.

(b) The velocity of the ray in the glass is from $n = c/v$, or

$$v = c/n = 3.0 \times 10^8 \text{ m/s}/1.62 = 1.85 \times 10^8 \text{ m/s}.$$

(c) Now that the velocities are known, the time difference requires finding the distance the rays travel. The distance a (the path through the slab) comes from $\cos 21^o = 2.0 \text{ cm}/a$, or $a = 2.14 \text{ cm}$. Construct a line perpendicular to the two rays at the point where the refracted one leaves the slab. The distance b in the original ray direction comes from $\cos 14^o = b/2.14 \text{ cm}$, or $b = 2.08 \text{ cm}$.

The time for the refracted beam is $\tau_r = 0.0214\,\text{m}/1.85 \times 10^8\,\text{m/s} = 11.6 \times 10^{-11}\,\text{s}$.

The time for the unaltered beam is $\tau = 0.0208\,\text{m}/3.0 \times 10^8\,\text{m/s} = 6.9 \times 10^{-11}\,\text{s}$.

The time delay is $4.7 \times 10^{-11}\,\text{s}$.

Geometry is very important for these types of problems. Here's a few items from geometry to remember:

1. When two lines intersect, alternate interior angles are the same

2. In any triangle, the interior angles must all add up to 180°. The triangle at right is an equilateral triangle, with all angles equal. If the bottom two angles were 70°, then the top angle would be 40°.

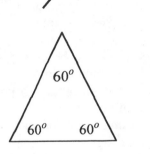

If you get bogged down trying to find an angle in a complex geometry, keep these two rules in mind. Draw your own triangles within the system and just start figuring out as many angles as you can.

22

MIRRORS AND LENSES

The subject of mirrors and lenses is difficult to treat broadly because for a thorough discussion there needs to be considerable attention to sign convention. Different instructors and authors approach the subject with slight but very significant differences that impact the working of problems.

DANGER

BEWARE OF FLOATING MINUS SIGNS

Watch Out!

The only way to become proficient in working with lenses and mirrors is to place the sign conventions appropriate to your course on a card in front of you and do problems drawing the rays and working the calculations. Be sure that you know the sign conventions and practice doing sample problems before taking any tests on mirrors or lenses. Signs will be your major (perhaps only) source of error. Sample sign conventions for mirrors and lenses are given below.

Sign Conventions for Mirrors

The object distance is the distance of the object from the surface of the mirror:

> o is positive if the object is in front of the mirror
> o is negative if the object is in back of the mirror
> (some texts call the object distance d_o and others call it s)

The image distance is the distance of the image from the surface of the mirror:

> i is positive if the image is in front of the mirror
> i is negative if the image is in back of the mirror
> (some texts call the image distance d_i and others call it s')

The focal point is the point where light rays converge from a mirror:

f is positive if the center of curvature is in front of the mirror
f is negative if the center of curvature is in back of the mirror
(almost everybody calls this f)
The sign conventions for f are the same for R, the radius of curvature.

The magnification of a lens is the number that you multiply the object size by to get the image size:

M is positive if the image is erect
M is negative if the image is inverted
(some texts call this m instead of M)

Sign Conventions for Lenses

o is positive if the object is in front of the lens
o is negative if the object is in back of the lens

i is positive if the image is in back of the lens
i is negative if the image is in front of the lens

f is positive for a converging lens
f is negative for a diverging lens

Mirrors

Flat Mirror

An object placed in front of a plane mirror appears to an observer to be behind the mirror.

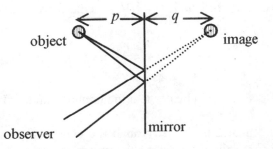

Because of the law of reflection, divergent rays intercepted by the observer on reflection from the mirror appear to come from behind the mirror. The object distance, p, is numerically equal to the image distance, q. The image is called a **virtual image** because the light does not physically come from the image. A **real image** is one where the light comes from or passes through the image.

22-1 An object is 4.0 cm in front of a plane (flat) mirror. Where does an observer looking into the mirror see the image and is it real or virtual?

Solution: The image is 4.0 cm behind the mirror. It is a virtual image because light does not pass through this image point.

Concave Mirror

A concave (converging) spherical mirror is shown below. Memorize the shape of a concave mirror and be sure that you can draw both a concave and convex mirror.

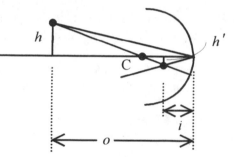

An object of height h placed at o, the object distance from the concave mirror, will produce a smaller image, h' at i, the image distance, according to the formula

$$\frac{1}{o}+\frac{1}{i}=\frac{2}{R}=\frac{1}{f}$$

where R is the radius of curvature and $f = (R/2)$ is the focal length. The magnification is

$$M=\frac{h'}{h}=-\frac{i}{o}=-\frac{R-i}{o-R}$$

The minus sign in the definition indicates that the image is inverted.

Draw a ray from the top of the object through the center of curvature (C in the figure above). Next draw a ray to the point where the principal axis (the horizontal line through C) intersects the mirror reflecting this ray back to intersect the one drawn previously. This ray just bounces straight back toward the object. The intersection of these two rays defines the top of the object. Drawing these rays requires experience. Set up several situations and draw the ray diagrams to become familiar with the procedure. Also, find out if you are going to need to draw ray diagrams for your test. If so, you need to practice doing this for different kinds of mirrors and lenses. If not, just concentrate on the sign conventions and formulas.

22-2 For a spherical concave mirror of 12 cm radius of curvature, describe the image of a 2.0 cm height object placed 20 cm on the center line of the mirror according to the figure above.

Solution: The image is inverted. The ray diagrams show this. Using the radius of curvature and the object distance, find the image distance from

$$\frac{1}{20\,\text{cm}} + \frac{1}{i} = \frac{2}{12\,\text{cm}} \quad \text{or} \quad i = 8.6\,\text{cm}$$

The magnification is from $M = -\dfrac{8.6}{20} = -\dfrac{12-8.6}{20-12} = -0.43$.

The height of the image is from $\dfrac{h'}{2.0\,\text{cm}} = -0.43$, or $h' = -0.86\,\text{cm}$.

The image is 8.6 cm from the mirror, inverted (minus sign) and real (rays pass through image).

When the object is at infinity (very far away) the mirror equation reduces to $1/i = 2/R$, and we can say that the rays from infinity are focused at $R/2$. This defines the focal length as $f = R/2$.

Convex Mirrors

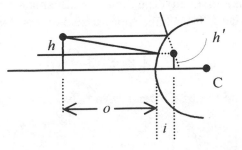

Objects placed in front of a convex mirror appear to come from behind the mirror, and they are smaller. First draw a ray from the top of the object parallel to the principal axis and reflect it from the mirror. This ray appears to come from the focus (behind the mirror). Next draw a ray so as to produce a reflected ray parallel to the principal axis of the mirror. The extension of this ray intersects the extension of the first one locating the top of the image.

The equations for concave mirrors also work for convex mirrors if a sign convention is adopted. Lengths to the left of the mirrors are positive, and lengths on the other side of the mirror (to the right of the mirror shown above) are negative. Lengths are measured (positive and negative) from the intersection of the principal axis and the mirror. These positive and negative regions are often referred to as the front and the back sides of the mirrors.

22-3 For a spherical convex mirror of 14 cm radius of curvature, describe the image of a 2.5 cm object placed 30 cm out on the principal axis of the mirror.

Solution: Here is where we get into the signs. The focus and the image are on one side of the mirror and the object is on the other side. Therefore, we take the focus as negative and expect the image distance to be negative. The image distance is

$$\frac{1}{30\,cm} + \frac{1}{i} = -\frac{1}{7\,cm} \quad \text{or} \quad i = -5.7\,cm$$

The magnification is $M = -\frac{-5.7}{30} = \frac{-14 + 5.7}{30 + 14} = 0.19$.

The height of the image is $\frac{h'}{2.5\,cm} = 0.19$, or $h' = 0.48\,cm$.

The minus sign for the image distance indicates that the image is behind the mirror, or on the same side as all the other minus signs. The plus sign for the magnification indicates that the image is erect (not inverted). The image is virtual since it is behind the mirror and no light is physically passing through the spot where the image appears to be.

Lenses

There are two types of thin lenses, converging and diverging:

converging   diverging

The converging lens converges parallel rays to a point called the focus, while a diverging lens refracts rays to make them appear as to come from a focus. The sign conventions become more involved for lenses than for mirrors. We will handle the signs in the context of each problem. The relationship between image distance, object distance, and focal length is the same as for mirrors.

22-4 A converging lens of focal length 8.0 cm forms an image of an object placed 20 cm in front of the lens. Describe the image.

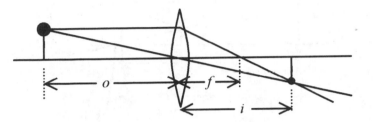

Solution: Draw a ray from the top of the object parallel to the axis then through the focus. Next draw a ray from the top of the object through the center of the lens to intersect the first ray. This locates the top of the image. Use the same equation we used with mirrors to find the image distance:

$$\frac{1}{o}+\frac{1}{i}=\frac{1}{f} \implies \frac{1}{20 \text{ cm}}+\frac{1}{i}=\frac{1}{8.0 \text{ cm}} \implies i=13 \text{ cm}$$

The magnification is $(-)$ image distance over object distance or $M = -\frac{i}{o} = -\frac{13}{20} = -0.67$.

The image is located 13cm on the side of the lens opposite the object with magnification 0.67. It is inverted (minus sign) and real (rays pass through image).

22-5 A diverging lens has a 14 cm focal length. Describe the image of a 4.0 cm object placed 40 cm from the lens.

Solution: Draw a ray from the top of the object to the lens parallel to the principal axis and refract it back to the focus. Next draw a line from the top of the object through the center of the lens. The intersection of these rays locates the top of the object. The image distance is

$$\frac{1}{40 \text{ cm}}+\frac{1}{i}=-\frac{1}{14 \text{ cm}} \text{ or } i=-10 \text{ cm}$$

The negative sign for the focal length is because this is a diverging lens. The image distance is negative because it is on the same side of the lens as the object (the opposite situation from a converging lens). The magnification is $10/40 = 0.25$.

The image is $0.25 \times 4.0 \text{ cm} = 1.0 \text{ cm}$ high, erect, virtual, and appears to come from a point 10 cm from the lens on the same side as the object.

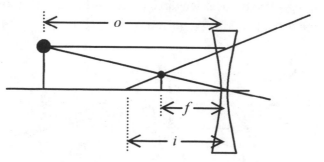

Go back over the problems in this chapter paying particular attention to the signs. As an exercise change the numbers in these problems and work them through until the sign conventions are clear in your mind.

There are a couple of variations on these types of problems. One is where you are given, for example, an image distance and focal length and asked to find the object distance. It is just a matter of using the same equations with the variables in different places. Another type of problem is multiple lens systems. If two lenses are placed in a sequence, first calculate the image distance and magnification for the first lens, and then that becomes the object for the second lens.

23

INTERFERENCE AND POLARIZATION

Things to Remember

- For **double** slit diffraction, the condition for constructive interference (or bright fringes) is $d\sin\theta = m\lambda$, where d is the distance between the two slits, λ is the wavelength of the light, and m is an integer $(0....1...2...3...$etc.$)$. The angles θ are where the light has maximum intensity.
- For **double** slit diffraction, the condition for destructive interference (dark fringes) is $d\sin\theta = (m+1/2)\lambda$.

- For **single** slit diffraction, the condition for destructive interference is $\sin\theta = \dfrac{m\lambda}{a}$, where a is the width of the single slit.
- When light enters a medium with a refractive index n, the speed of light is divided by n and the wavelength also gets divided by n. The frequency remains the same.

- Brewter's law: $\tan\theta_p = \dfrac{n_{refr}}{n_{inc}}$. This law is used to find the angle of incidence, θ_p, that causes the reflected and refracted rays to be perpendicular and thus completely polarized.

23-1 Light with a wavelength of 600 nm falls on a double slit diffraction setup and the second bright fringe is seen at an angle of 23°. (a) Find the distance between the slits. (b) Find the angle for the first bright fringe and (c) find the angle for the $m = 1$ dark fringe.

Solution: (a) $d\sin\theta = m\lambda \implies d\sin(23^o) = 2(600\,\text{nm}) \implies d = 3071\,\text{nm}$

(b) $3071\,\text{nm}\,\sin\theta = 1(600\,\text{nm}) \implies \sin\theta = 0.195 \implies \theta = \sin^{-1}(0.195) = 11.2^o$

(c) $d\sin\theta = (m+1/2)\lambda \implies 3071\,\text{nm}\,(\sin\theta) = 1.5(600\,\text{nm}) \implies \theta = 17^o$

The diagram below shows how the fringes are arranged. There is a bright center $m = 0$ fringe which is directly opposite the midpoint between the two slits. Next come the dark $m = 0$ fringes on either side, then the $m = 1$ bright fringes, then the $m = 1$ dark fringes, and so on.

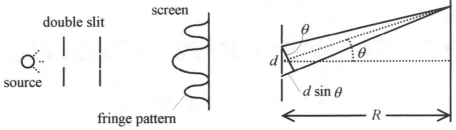

23-2 Waves are broadcast from a 1800 kHz radio station which is 1 km from a home. The radio transmitter is at the same height as the home. The home is in the path of an airport, and when planes fly over head, the signal is sometimes lost when a plane passes directly overhead. Find the first minimum height of the plane where the condition for destructive interference is satisfied at the home.

Solution: The key to doing this problem is having the understanding that destructive interference will happen when the two path lengths for the signal differ by half of one wavelength. The first path length is a direct line from the transmitter to the home. The second is the path to the plane and then reflected to the home ($c + b$ below). This is not a slit diffraction problem, just a destructive interference question. Let's get the value of ½ a wavelength first:

Insight

$$c = \upsilon\lambda \implies 3 \times 10^8 \, \text{m/s} = 1800 \times 10^3 /\text{s}(\lambda) \implies \lambda = 166.67 \, \text{m}$$

So destructive interference happens when the path lengths differ by 83.3 meters.

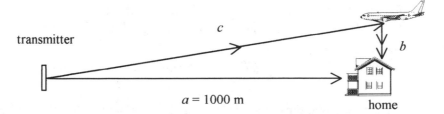

In the triangle drawn above, we need to take into account the geometry. The first path length is 1000 m, shown by a in the triangle above. The other path length is $c + b$. We know two things:

$$c^2 = a^2 + b^2 \implies c^2 = (1000\text{m})^2 + b^2$$

and
$$c + b = 1083.3 \text{ m}$$

Never underestimate the number of times the old two equations and two unknowns can be worked into a physics problem. This should be a piece of cake by now since you've been through Kirchhoff's law problems. The value of b is what we are looking for, so let's plug the bottom equation into the top one as follows:

$$(1083.3 - b)^2 = (1000)^2 + b^2$$

$$1083.3^2 - 2 \times 1083.3b + b^2 = 1000^2 + b^2$$

$$b = 80.1 \text{ m}$$

Go back and look at the geometry again. It should make sense to you that the answer is a little bit less than the 83.3 m that we got for the value of ½ wavelength (because c is a little bit larger than a).

23-3 Monochromatic light is incident on a single 4.0×10^{-4} m wide slit producing a diffraction pattern on a screen 1.5 m away. The distance from the central maxima to the second dark fringe is 3.4×10^{-3} m. What is the wavelength of the light?

Solution: The second dark fringe occurs when the path difference is 1 ½ wavelengths. For this question most texts invoke some mathematical trickery. Let's start with a diagram:

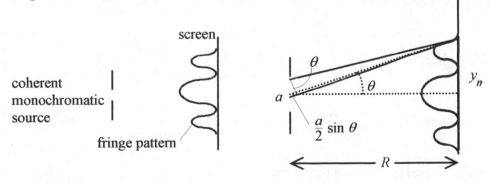

The y_m that we have drawn here is the distance along the screen between the center of the center bright fringe and the mth dark fringe. The basic equation for the single slit diffraction is $\sin\theta = \dfrac{m\lambda}{a}$. We don't have θ, however. There is a mathematical trick that involves the approximation that for small values of θ, $\sin\theta = \tan\theta$, and as a result we get: $\lambda = \dfrac{a\,y_m}{mR}$. For speed, forget the trick, just memorize this equation. Most texts don't list this equation as important, but it is essential unless you know the math trick.

Speed

$$\lambda = \frac{a\,y_m}{mR} = \frac{4.0\times10^{-4}\,\text{m}\cdot3.4\times10^{-3}\,\text{m}}{2\cdot1.5\,\text{m}} = 453\,\text{nm}$$

23-4 A grating with 2000 lines per cm is illuminated with a hydrogen gas discharge tube. Two of the hydrogen lines are at 410 and 434 nm. What is the first order spacing of these lines on a screen 1.0 m from the grating?

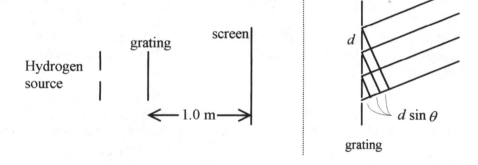

Solution: The phrase "first order" refers to the first maxima.

First find the spacing $1.0\times10^{-2}\,\text{m}/2000$ lines $= 5.0\times10^{-6}\,\text{m}$. Use $\lambda = \dfrac{d\,y_m}{mR}$.

For the 410 nm line, $y_1 = \dfrac{1\cdot\lambda}{d}\dfrac{R}{} = \lambda\dfrac{R}{d} = 410\times10^{-9}\,\text{m}\dfrac{1.0\,\text{m}}{5.0\times10^{-6}\,\text{m}} = 0.082\,\text{m}$.

For the 434 nm line, $y_1 = \dfrac{1\cdot\lambda}{d}\dfrac{R}{} = \lambda\dfrac{R}{d} = 434\times10^{-9}\,\text{m}\dfrac{1.0\,\text{m}}{5.0\times10^{-6}\,\text{m}} = 0.087\,\text{m}$.

The lines are separated by 5 mm on the screen.

Polarization

Light from a light bulb or the sun is circularly polarized. This means that if you could look at the electric field vectors, there would be no preferred direction. The electric field vectors would be randomly oriented in space. Polarization is described in terms of electric vectors rather than magnetic vectors. A radio transmitting antenna that is vertical produces waves with vertical electric vectors, and the radiation from this antenna is vertically polarized.

There are two ways to polarize light. It can be polarized by passing through a material that contains molecules all oriented in the same direction that absorb the electric vector (in this one direction) in the light. These materials are called **polaroids**. Polaroid sun glasses are made of a polaroid that absorbs the electric vector in one direction. Light also can be polarized by reflection at a certain angle.

The angle of incidence for complete polarization occurs when the angle between the reflected and refracted beams is 90°. The geometry of the situation is shown in the figure below.

The criterion for complete polarization is that $\theta_r + \theta_p = 90^o$. Also, $\tan\theta_p = \dfrac{n_{refr}}{n_{inc}}$.

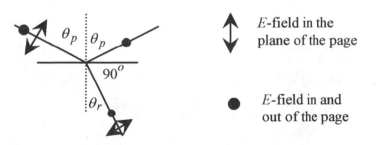

E-field in the plane of the page

E-field in and out of the page

This polarization law is known as **Brewster's law,** and the angle of polarization is called the **Brewster angle.** In working problems **be careful not to invert the fraction of the indexes of refraction.** The index for the material that the light reflects off of (and refracts into) goes on top. The material that the light is incident from goes on the bottom.

Watch Out!

23-5 What is the polarization angle for light reflected from water?

Solution:
$$\tan\theta_p = \frac{n_{refr}}{n_{inc}} = \frac{1.33}{1} \quad \text{or} \quad \theta_p = 53^o$$

23-6 A beam of unpolarized light in water is incident on what is reported to be diamond. By means of polaroids, the angle (of incidence) for complete polarization is determined to be 61°. Given that diamond should have an index of refraction of 1.81, is this actually diamond?

Solution: For light in water reflected from diamond, the Brewster angle should be

$$\tan \theta_p = \frac{n_{refr}}{n_{inc}} = \frac{2.42}{1.33} = 1.81 \quad \text{or} \quad \theta_p = 61^o$$

This is the correct index of refraction, so we conclude that this is diamond.

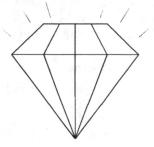

24

RELATIVITY

Things to Remember

- The first postulate of special relativity: "The laws of physics are the same in all inertial reference frames." Translation: You use the same  equations to describe velocity, acceleration etc.....or any physical law whether you are in an airplane flying at 300 miles per hour or standing on the ground. As long as you and the reference frame are all moving along together, this statement holds. After all, the earth is moving around the sun, the sun is moving with respect to other stars in the galaxy, our Milky Way galaxy is moving in the universe. It's all relative! Now if you drop a rock from the airplane that is going 300 miles per hour, your perception of how it falls is different than the person's who is standing still on the ground. That's because one reference frame is moving relative to the other. Really weird things start to happen if the motion of one reference frame with respect to the other gets close to the speed of light.

- The second postulate of special relativity: The speed of light is the same in all inertial frames. That means that if I am moving forward at half the speed of light and someone shines a flashlight at me from behind, I still measure the speed of light from the flashlight to be the same as if I were at rest. Keep this clear though: I will measure a different **wavelength** of the light, but not a different speed. If I am moving away from the source, I would measure a longer wavelength in accordance with the Doppler effect (this effect is discussed in the chapter on sound).

- Memorize this : $\gamma = \sqrt{1 - \dfrac{v^2}{c^2}}$. This factor is going to pop up a lot in relativity problems. In this formula, v is relative velocity that an object has and c is the speed of light. Remember this: The value of γ is always less than or equal to 1.

24-1 A spaceship is travelling at 90% of the speed of light with respect to the earth. Two signals are sent from the spacecraft to the earth. The people in the spacecraft send these two signals 9 hours apart from their reference frame. What is the time interval between the two signals as measured by people on the earth?

Solution: Use the equation $\Delta t = \dfrac{\Delta t_o}{\gamma}$. In this equation, Δt_o is the difference in time between two events for the observer in the same reference frame as the events. The observer sees the events happening in the same place. The signal is being sent by the spacecraft, so they are in the Δt_o reference frame. 90% of the speed of light is 0.90 c and $(0.90c)^2 / c^2 = 0.90^2$.

$$\gamma = \sqrt{1 - .9^2} = .436$$

$$\Delta t = \frac{9\,\text{hours}}{\gamma} \quad \Rightarrow \quad \Delta t = 20.6\,\text{hours}$$

24-2 A spaceship travels 50 light years at 0.9990 c. How long does the trip take in the reference frame of (a) an observer on the earth and (b) a passenger in the spaceship?

Solution: (a) Fifty years. Fifty light years just means the distance that light travels in 50 years. Keep in mind that the observer on earth always measures a longer time than the observer in the moving reference frame.

(b) $\gamma = \sqrt{1 - (0.999)^2} = 0.0447$. So the big question is, do we divide 50 by γ or multiply 50 by γ? Again, the t_0 is the reference frame of the spacecraft. So the answer is the time from the perspective of the spacecraft is:

$$50\,\text{years} \times 0.0447 = 2.24\,\text{years}$$

24-3 For the situation above, what is the distance in light years that the spaceship measures to the destination as they are travelling at the speed of 0.999 c?

Solution: For this we use the length contraction equation: L_0 here is the "proper length" between two points as measured by an observer at rest with respect to the two points. This is clearly the reference frame of the earth. So the answer is:

$$L = L_0 \gamma = 50\,\text{light years} \times 0.0447 = 2.24\,\text{light years}$$

Hopefully, this makes sense to you in the context of the previous problem. From the perspective of a person travelling near the speed of light, the whole trip makes perfect sense. They measure the distance they are travelling as 2.24 light years and since they are travelling at almost the speed of light, it takes them 2.24 years to get there.

24-4 You are in a railroad car moving at rest with respect to the surface of the earth. (a) What do you measure as the time it takes for a coin to drop 1.0 m to the floor of the railroad car? (b) What does an observer in another railroad car travelling at a constant 25 m/s measure for this time?

Solution: (a) Here we apply the equation of motion for a falling object from the first semester of physics:

$$x = v_0 t + (1/2)at^2$$

$$t = \sqrt{\frac{2x}{g}} = \sqrt{\frac{2 \cdot 1.0\ \text{m}}{9.8\ \text{m/s}^2}} = 0.45\ \text{s}$$

(b) Now here there is a shortcut that can save you time and some hassles with your calculator. First we want to get γ, which is

$$\gamma = \sqrt{1 - \frac{(25)^2}{(3 \times 10^8)^2}} = \sqrt{1 - 2.08 \times 10^{-14}} = ???!!!!!$$

Try this calculation on your calculator and see if you just keep getting 1.0000. The problem is many calculators don't go out to enough decimal places. OK, so the trick is this: For very small v/c, the best way to make this calculation is to use the following approximation:

Speed

$$\gamma = 1 - \frac{1}{2}\frac{v^2}{c^2} = 1 - \frac{1}{2}\left(\frac{25}{3.0 \times 10^8}\right)^2 = 1 - 3.5 \times 10^{-15}$$

You may want to memorize this little trick in case you get into this situation on a test. As far as the answer is concerned, this means that the correction is $\Delta t = \dfrac{\Delta t_0}{\gamma} = \dfrac{45\ \text{seconds}}{(1 - 3.5 \times 10^{-15})\ \text{s}}$. We'll just leave the answer in that form since this is a very minute correction.

24-5 What is the effective mass of an electron moving at $0.80\,c$?

Solution: There is another formula involving γ. This one involves mass. The formula

is $m = \dfrac{m_0}{\gamma}$. Here, m_0 is the rest mass and m is the mass moving at the velocity v.

$$m = \frac{m_0}{\sqrt{1 - v^2/c^2}} = \frac{9.1 \times 10^{-31}\,\text{kg}}{\sqrt{1 - 0.80^2}} = 15.2 \times 10^{-31}\,\text{kg}$$

24-6 Intergalactic space travellers need to know the relative velocities and masses of their space ships. Each ship, therefore, has a 1.0 m long bar painted on the side of the ship alongside their rest mass. As you pass by a ship, you measure this 1.0 m bar as 0.93 m. (a) What is your relative velocity? (b) You also observe their rest mass printed as 365,000 kg. What is their mass relative to you? (c) What does an observer in the other ship measure for your 1.0 m bar?

Solution: (a) The length you observe is $L = L_0 \sqrt{1 - v^2/c^2}$ where L_0 is the 1.0 m, the length an observer at rest with respect to the vehicle would measure, and L is the length you measure so

$$0.93 = 1.0 \quad \sqrt{1 - v^2/c^2} \quad \text{or} \quad v = 0.37\,c$$

(b) The relativistic mass you observe is

$$m = \frac{m_0}{\sqrt{1 - v^2/c^2}} = \frac{365,000\,\text{kg}}{0.93} = 392,000\,\text{kg}$$

(c) Observers in the other space ship also measure your bar as 0.93 m and relative speed as 0.37 c. You have the same relative speed to them as they have to you.

25

QUANTUM PHYSICS

Things to Remember

- Light can be thought of as particles, each having an energy of $E = hf$. f is the frequency and h is Plank's constant, $h = 6.63 \times 10^{-34}$ Joules·seconds. The dual nature of light can be confusing. Sometimes it behaves like a wave, sometimes like particles which are called photons.

- For questions that ask about the work function or photoelectric effect, memorize the equation $hf = \phi + (1/2)m_e v^2$. See problem 25-3. Remember the conversion between Joules and electron volts: $1\text{eV} = 1.6 \times 10^{-19}$ J.

- The momentum of a photon of light is $p = \dfrac{h}{\lambda}$. This equation can also be used for calculating the de Broglie wavelength of particles.

- When a photon bounces off an electron, it is called Compton scattering. The equation that relates changes in wavelength of the photon to the scattering angle is $\Delta\lambda = \lambda' - \lambda = \dfrac{h}{mc}(1 - \cos\theta)$. For speed on a test you can memorize that $h/mc = 2.43 \times 10^{-12}$ m.

Speed

- Heisenberg's uncertainty principle states that you cannot know the position and velocity of an object with absolute certainty. In equation form this is written $\Delta x \Delta p \geq \dfrac{h}{2\pi}$. A similar relation holds for energy and time, $\Delta E \Delta t \geq \dfrac{h}{2\pi}$.

25-1 A beam of light consists of photons with an energy of 5.5×10^{-19} J. What is the wavelength and frequency of the light (a) in air, and (b) if it is travelling in glass with an index of refraction of 1.33?

Solution: (a) $E = hf \implies 5.5 \times 10^{-19}\,\text{J} = (6.63 \times 10^{-34}\,\text{J}\cdot\text{s})f \implies f = 8.30 \times 10^{14}\,\text{Hz}$

To get wavelength use

$$c = \lambda f \implies 3 \times 10^8 \, \text{m/s} = \lambda (8.30 \times 10^{14} \, \text{Hz}) \implies \lambda = 3.61 \times 10^{-7} \, \text{m} = 361 \, \text{nm}$$

(b) Remember that the frequency of light does not change between glass and air. Frequency is always constant and so is photon energy. The index of refraction of 1.33 means that the speed gets divided by 1.33 and the wavelength gets divided by 1.33. Therefore, $\lambda_{glass} = 271 \, \text{nm}$.

25-2 A radio station broadcasts at 88.3 MHz. If the power is 50 Megawatts, how many photons per second are emitted from the antenna?

Solution: The energy of each individual photon is

$$E = 6.63 \times 10^{-34} \, \text{J} \cdot \text{s} (88.3 \times 10^6 \, \text{Hz}) = 5.85 \times 10^{-26} \, \text{Joules/photon}$$

Fifty Megawatts means 50×10^6 Joules/second . So we can find the number of photons per second by

$$\frac{50 \times 10^6 \, \text{Joules}}{} \times \frac{1 \, \text{photon}}{5.85 \times 10^{-26} \, \text{Joules}} = 8.55 \times 10^{32} \, \text{photons}$$

25-3 Radiation with a wavelength of 200 nm strikes a metal surface in a vacuum. Ejected electrons have a maximum speed of 1×10^6 m/s. What is the work function of the metal in eV?

Solution: This is a photoelectric effect problem. The idea here is that each photon of light has a certain amount of energy. That energy **goesinto** two places. First, it goes into pulling the electron out of the metal (this is the energy of the work function which is labeled ϕ). The second place the original energy goes is into kinetic energy of the ejected electron. This can be written as follows.

$$hf = \phi + (1/2)m_e v^2$$

First, convert 200 nm to frequency:

$$c = \lambda f \implies 3 \times 10^8 \, \text{m/s} = 200 \times 10^{-9} \, \text{m}(f) \quad f = 1.5 \times 10^{15} \, \text{Hz}$$

$$(6.63 \times 10^{-34} \, \text{J} \cdot \text{s}) 1.5 \times 10^{15} \, \text{Hz} = \phi + (1/2)(9.1 \times 10^{-31} \, \text{kg})(1 \times 10^6 \, \text{m/s})^2$$

$$9.94\times10^{-19}\,\text{J} - 4.55\times10^{-19}\,\text{J} = \phi$$

$$\phi = 5.49\times10^{-19}\,\text{J}\times\frac{1\text{eV}}{1.6\times10^{-19}\,\text{J}} = 3.43\,\text{eV}$$

25-4 Find the momentum of a photon with a wavelength of 200 nm.

Solution: $$p = \frac{h}{\lambda} = \frac{6.63\times10^{-34}\,\text{J}\cdot\text{s}}{200\times10^{-9}\,\text{m}} = 3.32\times10^{-27}\,\text{kg}\cdot\text{m/s}$$

The units of momentum come out to kg·m/s which fits the basic definition of momentum from the first semester of physics which was momentum = mass × velocity .

25-5 What is the change in wavelength of an X-ray photon when it scatters from an electron at an angle of 40° ?

Solution: The analysis here is that a photon comes in with a certain amount of momentum before the collision. After the collision, the electron has increased momentum and the photon has a reduced momentum. This means its wavelength changes. This is called Compton scattering. The formula that we use is

$$\Delta\lambda = \lambda' - \lambda = \frac{h}{mc}(1 - \cos\theta)$$

$$\Delta\lambda = 2.43\times10^{-12}\,\text{m}(1 - \cos40^o) = 5.69\times10^{-13}\,\text{m}$$

25-6 Calculate (a) the de Broglie wavelength of an electron with kinetic energy of 100 keV, (b) an electron with kinetic energy of 1 eV, and (c) a golf ball with a mass of 45 grams and a velocity of 50 meters per second.

Solution: Just as light can be described either as a particle or a wave, so alos particles can be described as either particles or waves. To find the wavelength of a particle, use the same formula as we did for light: $p = \dfrac{h}{\lambda}$.

(a) We are given kinetic energy of an electron, but we need the momentum.

$$100 \times 10^3 \, \text{eV} \frac{1.6 \times 10^{-19} \, \text{J}}{\text{eV}} = 1.6 \times 10^{-14} \, \text{J} = \frac{1}{2} m_e v^2 = \frac{1}{2}(9.1 \times 10^{-31} \, \text{kg})v^2$$

$$v = 1.88 \times 10^8 \, \text{m/s}$$

$$p = mv = 9.1 \times 10^{-31}(1.88 \times 10^8 \, \text{m/s}) = \frac{h}{\lambda} = \frac{6.63 \times 10^{-34} \, \text{J} \cdot \text{s}}{\lambda}$$

$$\lambda = 3.9 \times 10^{-12} \, \text{m}$$

(b) $$1 \, \text{eV} \frac{1.6 \times 10^{-19} \, \text{J}}{\text{eV}} = 1.6 \times 10^{-19} \, \text{J} = \frac{1}{2} m_e v^2 = \frac{1}{2}(9.1 \times 10^{-31} \, \text{kg})v^2$$

$$v = 5.93 \times 10^5 \, \text{m/s}$$

$$p = mv = 9.1 \times 10^{-31}(5.93 \times 10^5 \, \text{m/s}) = \frac{6.63 \times 10^{-34} \, \text{J} \cdot \text{s}}{\lambda}$$

$$\lambda = 1.22 \times 10^{-9} \, \text{m} = 1.22 \, \text{nm}$$

(c) For the golf ball, $$p = mv = .045 \, \text{kg}(50 \, \text{m/s}) = \frac{h}{\lambda} = \frac{6.63 \times 10^{-34} \, \text{J} \cdot \text{s}}{\lambda}$$

$$\lambda = 2.95 \times 10^{-34} \, \text{m}$$

The wave nature of particles only really becomes an observable effect for very small particles like electrons. Since the distance between atoms in a crystal is usually a few tenths of 1 nm, we can see that the de Broglie wavelength of electrons can be approximately the same, depending on their velocity. This means that wave effects such as interference and diffraction will be observed when electrons interact with solids.

25-7 The velocity of a proton is $3.5 \times 10^7 \, \text{m/s} \pm 1.5\%$. What is the uncertainty in its position?

Solution: The Heisenberg uncertainty rule for this is $\Delta x \Delta p \geq \dfrac{h}{2\pi}$. So first we need to get Δp, which is 3% of the momentum of the proton. ($\pm 1.5\%$ means that the velocity could be 1.5% higher or 1.5% lower than we think it is, so the uncertainty is 3%.)

$$\Delta p = 1.7 \times 10^{-27} \, \text{kg}(0.03)3.5 \times 10^7 \, \text{m/s} = 1.79 \times 10^{-21} \, \text{kg} \cdot \text{m/s}$$

$$\Delta x (1.79 \times 10^{-21} \, \text{kg} \cdot \text{m/s}) \geq \frac{6.63 \times 10^{-34} \, \text{J} \cdot \text{s}}{2\pi}$$

$$\Delta x = 5.91 \times 10^{-14} \, \text{m}$$

Perhaps it is only fitting that the last problem in this book be about the uncertainty principle. If you are preparing for your final exam and are uncertain about how to best prepare, please review the *Advice to the Utterly Confused*. Best of luck!

ABOUT THE AUTHORS

Dr. Daniel Oman (son) received the B.S. degree in physics from Eckerd College. He received both his M.S. degree in physics and his Ph.D. in electrical engineering from the University of South Florida. As a teaching assistant from 1989 to 1993, he taught many *utterly confused* students both in classes and one-on-one. He has done research on CO_2 lasers and solar cells, and he has authored several technical articles. Dan is currently a Member of Technical Staff at Lucent Technologies, where he works on process integration in the manufacturing of microelectronic circuits.

Dr. Robert Oman (father) received the B.S. degree from Northeastern University and the Sc.M. and Ph.D. degrees from Brown University, all in physics. He has taught mathematics and physics at several colleges and universities including University of Minnesota, North Shore Community College, Northeastern University, and University of South Florida. He has also done research for Litton Industries, United Technologies, and NASA, where he developed the theoretical model for the first pressure gauge sent to the moon. He is author of numerous technical articles, books, and how-to-study books, tapes, and videos.

Visit the authors' website at www.rdoman.com
Comments and suggestions about the book are welcomed.